1st International Conference on Advanced Engineering Functional Materials (ICAEFM 2018)

IOP Conference Series: Materials Science and Engineering
Volume 410

Bhubaneswar, India
21 – 23 September 2018

ISBN: 978-1-5108-7195-3
ISSN: 1757-8981

Printed from e-media with permission by:

Curran Associates, Inc.
57 Morehouse Lane
Red Hook, NY 12571

Some format issues inherent in the e-media version may also appear in this print version.

Printed by Curran Associates, Inc. (2018)

For permission requests, please contact the Institute of Physics at the address below.

Institute of Physics
Dirac House, Temple Back
Bristol BS1 6BE UK

Phone: 44 1 17 929 7481
Fax: 44 1 17 920 0979

techtracking@iop.org

Additional copies of this publication are available from:

Curran Associates, Inc.
57 Morehouse Lane
Red Hook, NY 12571 USA
Phone: 845-758-0400
Fax: 845-758-2633
Email: curran@proceedings.com
Web: www.proceedings.com

Table of contents

Volume 410

2018

1st International Conference on Advanced Engineering Functional Materials (ICAEFM)
21–23 September 2018, GITA, Bhubaneswar, Odisha, India

Accepted papers received: 15 August 2018
Published online: 24 September 2018

Preface

Papers

IOP Conf. Series: Materials Science and Engineering **410** (2018) 011001 doi:10.1088/1757-899X/410/1/011001

Materials Science and Engineering has provided plenty of scope for research and development. It is no wonder why many national and international conferences are held globally and that the results of this several publications are made in many peer-reviewed Journals. This is due to increasing demand for newer advanced functional materials for every field of engineering. With an accelerated demand for communication, space technology, Nano engineering, bioengineering, etc. Many reports are brought out in the form of reports published by each country. It is prudent to organize the International Conference on Advanced Engineering Functional Materials covering different materials. The contribution by various scientists/researchers made the conference lively. The proceedings aim to publish their presentations for circulation among in rest of the world.

This may provide opportunities to new researchers to add to their knowledge.

IOP Conf. Series: Materials Science and Engineering **410** (2018) 011002 doi:10.1088/1757-899X/410/1/011002

Peer review statement

All papers published in this volume of *IOP Conference Series: Materials Science and Engineering* have been peer reviewed through processes administered by the proceedings Editors. Reviews were conducted by expert referees to the professional and scientific standards expected of a proceedings journal published by IOP Publishing.

Mechanical and biodegradation analysis of thermoplastic starch reinforced nano-biocomposites

Ajaya K. Behera[1]

[1]Department of Chemistry, Utkal University, Bhubaneswar 751004, India

Abstract

The aim of the present investigation is to develop and characterize nano-filler reinforced thermoplastic starch (TPS) composites. Both montmorillonite (MMT), and Cloisite 30B are used in different weight percentages as reinforcing filler with TPS by hot compression molding to develop starch-mmt (SM) and starch-cloisite (SC) composites respectively. Fabricated composites were mechanically tested. Composite containing 3 wt% mmt showed tensile strength and tensile modulus of 2.67 MPa and 68.4 MPa respectively X-ray diffraction (XRD) and transmission electron microscopy (TEM) of nano-composites were analyzed. Biodegradation of composites were carried out in soil burial condition for certain time periods and found that the degradation rate of composites are slow and lost only on an average of 45% of their original weight after 60 days under soil burial condition. Surface topography of degraded samples was studied under scanning electron microscope. As SM and SC composites are biodegradable in nature, these can be utilized as packaging materials, molded articles and other aesthetic products to reduce the use of non degradable synthetic plastic.

Key words: Thermoplastic starch; MMT; Cloisite 30B; Nano-biocomposites

1. Introduction

Packaging with petrochemical based plastics such as polyester, polyolefins, polyamides and polyolefin halides etc. have been used largely because of their abundance and low cost. Vegetable, grocery, food packaging, etc. are growing from day to day and the materials used for packaging are concerns for preservation of all types of foods and protection from oxidative and microbial spoilage. Packaging material can be prepared from synthetic and biodegradable plastic. Better aesthetic values, durability, mechanical properties are the prime factors to choose

IOP Conf. Series: Materials Science and Engineering **410** (2018) 012001 doi:10.1088/1757-899X/410/1/012001

synthetic polymer and their composites over biodegradable polymer. But these polymers are non-biodegradable and after use creates environment, soil, water and their incineration causes air pollution. So, plastic waste management is now a great challenge due to non availability of free land for solid waste disposal [1-2].

Research on biodegradable polymer and their composite films has been increased now a day to counter the above problem. Biopolymers like starch, protein, chitosan, and cellulose are the best choices for researcher as they are renewable, cheap and abundant source. Scientifically these molecules are polar and chemically active in nature and completely biodegradable. Starch, a polymeric carbohydrate consisting of a large number of glucose units joined by glycosidic bonds and contains a number of hydroxyl groups [3-4]. Muller et al. (2012) prepared cassava starch reinforced natural and organically modified nano composites and found the use of 5% hygroscopic nanoclay increases strength and decreases the permeability to water vapor [5]. Priyada et al. (2013) studied mechanical, thermal and structural properties of rice starch films reinforced with rice starch nanocrystals and found the rice starch films containing 20% starch nanocrystals had the best mechanical properties and provided a high tensile strength and elongation at break of about 16.43 MPa and 5.76%, respectively [6]. Slavutsky et al. (2012) reported on water barrier properties of corn starch-clay nanocomposite films that addition of nanoparticles at percentages below 10% had excellent potential for their application in the technology of biopolymer based films [7].

Park et al. (2002) studied moisture and mechanical properties of starch/nanoclay composites employing an organically modified MMT and natural MMT. They reported that modified MMT reinforced composite showed higher water vapor transmission rates than samples with MMT [8]. Mohan et al. (2016) studied the effect of nanoclay fillers on the properties of corn starch polymer-based biofilm and the results reported are reduction of 22% in water absorption, 40% in moisture uptake, 30% in oxygen permeation and 31% in swelling for 2–3 wt % nanoclay-filled biofilm as compared with unfilled biopolymer [9]. Hassani and Nafchi (2014) prepared and characterized potato starch nanoclay composite and found increase in tensile strength from 7.33 to 9.82 MPa, and decrease in elongation at break from 68.0 to 44.0% [10].

Biodegradable thermoplastic starch (TPS) reinforced nano composites are a new class of hybrid engineering compounds which provides better mechanical properties, dimensional stability, water resistance and biodegradability over non-degradable synthetic polymer [11]. In this work biodegradable polyester blended nano-composites were developed and characterized. Both natural nanoclay (montmorillonite (MMT)) and organically modified nanoclay (Cloisite 30B) of different weight percentages were utilized separately for fabrication of nano-composites. Developed bio-composites can be utilized in different sectors like packaging, decorating and automobile, etc.

2. MATERIALS AND METHODS

2.1. Materials

Thrmoplastic starch (TPS) (derived from corn starch and blended with polyester) was supplied by Bio-grade (Nanjing) Pty Ltd., China. Southern Clay Co. USA, supplied Montmorillonite (MMT) and Cloisite 30B. Glycerol (Merck, India) was procured from local market.

2.2. Fabrications of nanoclay reinforced TPS composites

Blends with various amounts (0,1, 2, 3,4 and 5 wt%) of natural nanoclay MMT, glycerol (5 wt%) and TPS were prepared by melt mixing at 100 ºC and 60 rpm for 5 min using an internal mixer having double screw extruder (Brabender mixing chamber). These different sets of TPS-MMT blends were collected and then compressed in a hot press at the temperature of 105 ºC for 10 min under 6.0 MPa pressures to prepare starch-mmt (SM) composites. Fabricated composites comprising 0-5 wt% of nanoclay are coded as SM0-5 respectively. Similarly TPS-Cloisite 30B (1, 2, 3, 4 and 5 wt%) composites (SC1-5) were prepared by following above formulation and procedure.

2.3. Characterizations

Tensile properties of SM and SC composites were characterized according to standard ASTM D638 with cross head speed of 5 mm/min using HOUNSFIELD H10K UTM instrument. From each of the composite eight specimens were tested and average value was reported. From both the sets, highest tensile strength shown composite was considered as mechanically optimized.

X-ray diffraction (XRD) study of optimized SM and SC composites was performed using a X-ray diffractometer (WAXD, ULTIMA-III, Rigaku, Japan) with nickel filtered Cu-Kα radiation (λ= 0.154nm) operated at 40 kV and 100 mA, at a scanning rate of 1 °/min.

Transmission electron microscope (TEM) analysis for bulk morphology study of optimized SM and SC composites were characterized using a transmission electron microscope, model JEM-1230, JEOL with an acceleration voltage of 100 kV.

Biodegradation analysis of optimized SM and SC were carried out under soil burial condition in accordance to the method specified in standard BIS 1623-1992. Field emission scanning electron micrograph (FE-SEM) analysis of optimized composites before and after degradation was taken by using a scanning electron microscope (SUPRA-40, Germany) instrument operated at an accelerating voltage of 5 kV.

3. RESULTS AND DISCUSSION

3.1. Mechanical strength analysis

Tensile strength, modulus, elongation at break values of SM and SC composites are reported in Table 1. With increase in mmt content from 0 to 5 wt%, tensile strength and tensile modulus of SM composites increased due to reinforcement of nanofiller. SM3 showed tensile strength and tensile modulus of 2.96 MPa and 66.4 MPa respectively. Elongation at break value of SM composites decreased with increase in nanoclay amount due to improvement in brittleness property [12]. Similar result obtained in case of SC composites. Tensile strength of cloisite 30B content composites enhanced from 2.54 MPa to 3.03 MPa. Tensile modulus of SC composites also increased from 58.7 MPa to 65.5 MPa. SC3 showed highest tensile strength and modulus among all SC composites and SM3 showed highest tensile properties among all SM composites, hence both are considered as mechanically optimized.

Table 1. Mechanical properties of SM and SC composites

Composite	Tensile strength (MPa)	Standard deviation (±)	Tensile modulus (MPa)	Standard deviation (±)	Elongation at break (%)	Standard deviation (±)
SM0	1.57	0.13	56.2	1.42	6.54	0.04
SM1	2.43	0.15	58.4	1.42	6.26	0.04
SM2	2.67	0.12	61.3	1.42	6.08	0.04
SM3	2.96	0.11	66.4	1.42	5.76	0.04
SM4	2.29	0.18	55.7	1.42	5.58	0.04
SM5	1.91	0.12	51.8	1.42	5.26	0.04
SC1	2.54	0.16	58.7	1.42	6.76	0.04
SC2	2.72	0.14	60.3	1.42	6.18	0.04
SC3	3.03	0.13	65.5	1.42	5.93	0.04
SC4	2.43	0.12	55.8	1.42	5.53	0.04
SC5	2.02	0.12	50.8	1.42	5.24	0.04

3.2. XRD analysis of optimized TPSM and TPSC

XRD graph of mmt, cloisite 30B, SM3, and SC3 is shown in Fig.1. Cloisite 30B nanoclay showed its characteristics peak at around 4.96° (2θ) which corresponds to interlayer clay spacing (d spacing) of 17.7 Å, consistent with the reported value [13]. In its concerned composite i.e.,

IOP Conf. Series: Materials Science and Engineering **410** (2018) 012001 doi:10.1088/1757-899X/410/1/012001

SC3 have no peak suggested exfoliation nature of nanoclay at the interphase. MMT showed its characteristic peak at around 7.17° (2θ) which is shifted to words lower angle 5.52° for SM3 composite stating intercalation of nanoclay in TPS [14].

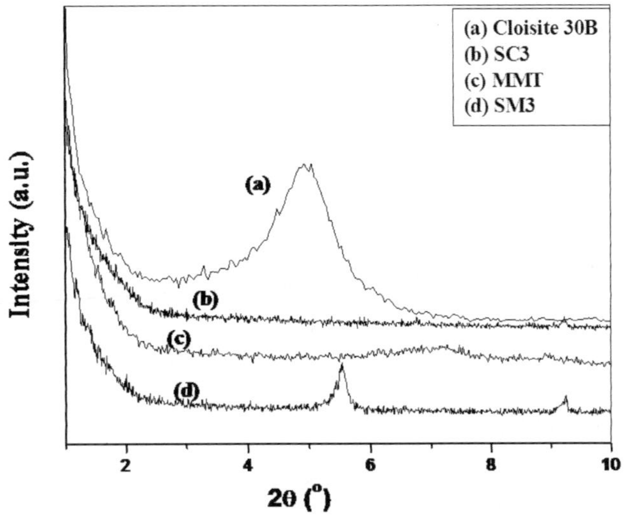

Fig.1. XRD graphs of mechanically optimized SM and SC composites

3.3. Morphology analysis of optimized SM and SC composites

TEM photographs of SM0, SM3, and SC3 is given in Fig.2. In SM0, smooth surface of thermoplastic starch is shown while in SM3 clay layers are clearly visible (marked by circle), arranged layer wise indicating intercalation of silicate layers as discussed in XRD. In SC3 composite, individual clay layers are separated from each other (marked by arrow) showing exfoliation of nanoclay in TPS. Both intercalation and exfoliation of nanoclay helps to improve the mechanical properties of TPS-nanoclay composite [14].

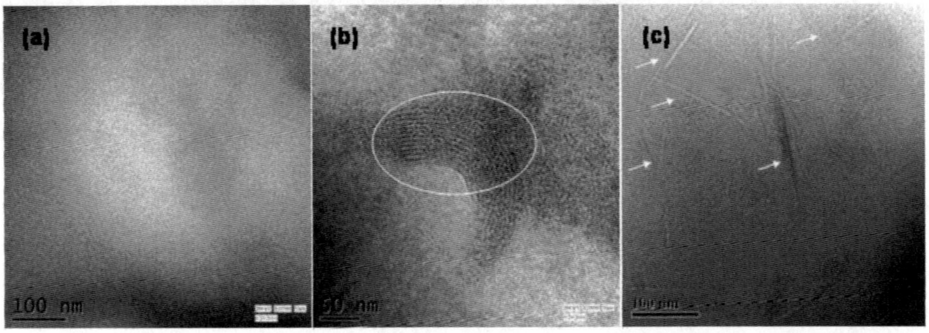

Fig.2. TEM photographs of (a) SM0, (b) SM3, and (c) SC3

3.4. Soil-burial degradation analysis of optimized SC and SM composites

Weight loss after different periods in soil burial of SM0, SM3, and SC3 are reported in Table 3. It was found SM0 without any nanoclay is less degraded initially as compared to SM3 and SC3. After 60 days under soil burial degradation SM0, SM3, and SC3 loosed 40.1%, 49.1% and 44.9% in weight respectively.

Table 2. Weight loss of SM0, SM 3, and SC3 after different biodegradation periods

Composite	After 15 days (%)	After 30 days (%)	After 45 days (%)	After 60 days (%)
SM0	9.2	12.6	28.7	40.1
SM3	14.9	18.2	34.7	49.1
SC3	10.2	14.1	31.2	44.9

3.5. FE-SEM analysis of biodegraded sample

Field-emission SEM photographs of biodegraded samples are given in Fig.3. SM3 (Fig.3(a)) and SC3 (Fig.3(c)) shows smooth surface before biodegradation. After 60 days of degradation it was found degraded SM3 (Fig.3(b)) and degraded SC3 (Fig.3(d)) has many cavities on its surface (marked by circle). Both composite surfaces were found roughed, degraded, and in maximum places some matrix has been eaten by microbes. This indicates the degradation of TPS-nanoclay composites [15].

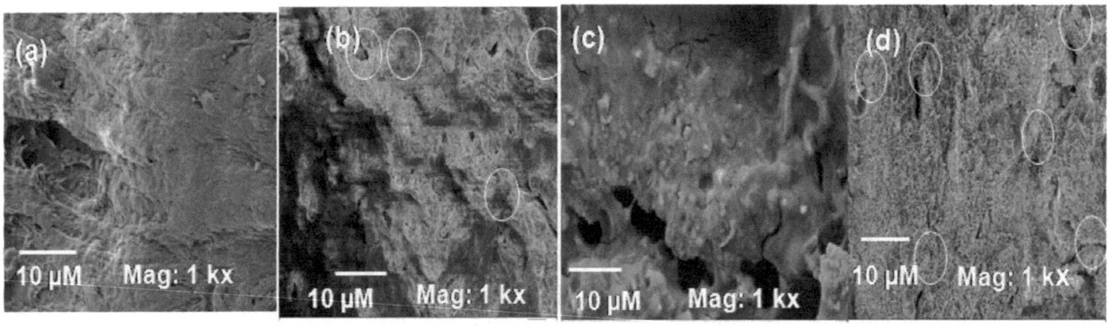

Fig.3. FE-SEM photographs of (a) SM3, (b) Degraded SM3, (c) SC3, and (d) Degraded SC3

4. CONCLUSIONS

The objective of this work was to develop and characterize biodegradable nano-bio composites. It was found with reinforcement of natural nanoclay tensile strength of composites increased upto 89%. Similarly 3 wt% organically modified nanoclay loaded composite (SC3) showed maximum tensile strength of 3.03 MPa which is nearly 93% enhancement. Both XRD

IOP Conf. Series: Materials Science and Engineering **410** (2018) 012001 doi:10.1088/1757-899X/410/1/012001

and TEM analysis of optimized composites proved formation of intercalation and exfoliation structure. Weight loss and FE-SEM photographs of degraded sample proved developed composites are biodegradable in nature. Hence these composites are eco-friendly unlike synthetic plastic and can be utilized in different sectors like packaging, decorating and automobile, etc.

References

[1] A.K. Behera, S.Avancha, S. Manna, R. Sen, and B. Adhikari, 2014, *Polymer Engineering and Science*, Vol. *54*, pp. 345–354.

[2] A.K. Behera, S. Avancha, R. Sen, and B. Adhikari, 2013, *Journal of Applied Polymer Science*, Vol. *12. No. 7*(6), pp. 4681-4687.

[3] A.K. Behera, S. Avancha, R.K. Basak, R. Sen, and B. Adhikari, 2012, *Carbohydrate Polymers*, Vol. *88*, pp. 329–335.

[4] C.M. Müller, J.B. Laurindo, and F. Yamashita, 2012, *Carbohydrate polymers*, Vol. *89. No.* 2, pp. 504-510.

[5] K. Piyada, S. Waranyou, and W. Thawien, 2013, *International Food Research Journal*, Vol. *20. No.* 1.

[6] A.M. Slavutsky, M.A. Bertuzzi, and M. Armada, 2012, *Brazilian Journal of Food Technology*, Vol. *15. No.* 3, pp. 208-218.

[7] Park, H.M., Li, X., Jin, C. Z., Park, C. Y., Cho, W. J., & Ha, C. S. (2002). *Macromolecular Materials and Engineering*, *287*(8), 553-558.

[8] T.P. Mohan, K. Devchand, and K. Kanny, 2017, *Journal of Plastic Film & Sheeting*, Vol. *33. No.* 3, pp. 309-336.

[9] F. Sadegh-Hassani, and A.M. Nafchi, 2014, *International journal of biological macromolecules*, Vol. *67*, pp. 458-462.

[10] M. Huang, J. Yu, and X. Ma, 2006, *Carbohydrate Polymers*, Vol. *63. No.* 3, pp393-399.

[11] V.P. Cyras, L.B. Manfredi, M.T. Ton-That, and A. Vázquez, 2008, *Carbohydrate Polymers*, Vol. *73. No.* 1, pp. 55-63.

[12] Q.X. Zhang, Z.Z. Yu, X.L. Xie, K. Naito, and Y. Kagawa, 2007, *Polymer*, Vol. *48. No.* 24, pp. 7193-7200.

[13] K. Okada, T. Mitsunaga, and Y. Nagase, 2003, *Korea-Australia Rheology Journal*, Vol. *15*. No. 1, pp. 43-50.

[14] E. Corradini, A.J.F.D. Carvalho, A.A.D.S. Curvelo, J.A.M. Agnelli, and L.H.C. Mattoso, 2007, *Materials Research*, Vol. *10No.* 3, pp. 227-231.

An advanced cured high carbon ferrochrome slag (HCFCS) geopolymer (GP): A constructional materials

Muktikanta Panigrahi[1], Saralasrita Mohanty[2], Ratan Indu Ganguly[1], Radha Raman Dash[1]

[1]Metallurgical Engineering Department, Gandhi Institute of Engineering and Technology, Gunupur, Rayagada, Odisha, India-765022
[2]School of Physical Sciences, National Institute of Science Education and Research, HBNI, Jatni, India

Abstract

In the recent past, investigations are reported on Geopolymer concrete. Geopolymer is known as inorganic polymer. Such materials are useful materials which are capable of replacing constructional materials such as conventional cement. In present investigations, Geopolymer is made from chromium bearing slag obtained from a ferrochrome industry (BRG steel). The mechanism involved is that the silicon and aluminium present in the high carbon ferrochrome (HCFC) slag reacts with alkali liquid, forming Geopolymer which binds other non-reactive materials in the slag. The highest strength level achieved for as-prepared Geopolymers are 11 and 15 MPa by curing 7 days and 28 days, respectively. X-ray diffraction patterns reveal that there is a gradual transformation from crystalline phase to non-crystalline phase which is due to polymeric transformation with increase in curing time. SEM micrographs confirm formation of glassy phases at a higher curing time which also corroborate with mechanical properties such as compressive strength. Distinct changes are observed in FTIR spectra *i.e*, increase in peak height as well as appearance of many other peaks if compared with FTIR spectrum of virgin material. This strength is due to polymeric reactions and formation of chains with the monomeric structure. Striking features of the TGA pattern is observed for the two materials *i.e.*, uncured and cured samples. DSC isotherms show oozing out of inbuilt water which has accumulated during condensation polymerization reaction. The strength level achieved for optimum combination of variable is found to be comparable to that of standard motar of grade (M15) as is used for constructional purpose.

Key words: Geopolymer, Slag, Cement and Compressive strength

Corresponding Author: muktikanta2@gmail.com

IOP Conf. Series: Materials Science and Engineering **410** (2018) 012002 doi:10.1088/1757-899X/410/1/012002

1. Introduction

Portland cement (PC) is a common structural material, widely used for construction purposes [1,2]. However, its production causes environmental problems owing to release of CO_2 in the atmosphere and there by polluting our universe [3]. The production of PC is extremely energy intensive process. Therefore, many researchers have used FA as resource material for production of Geopolymer, which can replace PC in the long run [1,3]. It is no wander why researches are being carried out for the replacement of PC with new generation material, *i.e.*, GP. As a result of this GP technology has emerged out as an alternative route or means for replacement of PC [4].

GPs are described to be a chain of network of inorganic molecules [4]. In the year of 2008, Davidovits has described GP to be a future constructional material which is predicted to replace PC in concrete construction [5]. The material can be prepared from FA or slag, which essentially contains silicon and aluminum bearing oxides [4]. These oxides are polymerized with chain of nucleus by the action of active alkali metal solution such as sodium or potassium [4]. Thus, GP material can now be considered to be an innovative material of 21 century [6]. Many researchers have described the polymer structure, consisting of Silicon-oxygen-Aluminium back-bone [4]. The advantage of GP lies in a fact that it is cheaper, unhazardous, fire resistance and environmental friendly materials [4]. There exist some reports on the use of micro silica in ferrochrome slag for the preparation of conventional low castable cements [7,8]. Ferrochrome slag essentially is procured from stainless steel industries.

In the present investigation, a slag is chosen from a ferrochrome industry. The slag contains ingredients such as Al_2O_3, SiO_2, MgO, CaO, Fe_2O_3, Cr_2O_3, etc [9]. Chromium present in

IOP Conf. Series: Materials Science and Engineering **410** (2018) 012002 doi:10.1088/1757-899X/410/1/012002

the slag is in the form of chromium oxide which is around 4-5% [10]. The slag also contains high carbon since it is produced during carbothermal reduction of chromium oxide in an arc furnace. There is a disposal problem of this material, experienced by the industry and hence it is thought better to utilize it for the production of a use material such as GP. Attempt is also made to optimize process parameters for maximizing strength properties. Wherever, required detail investigations are carried out to explain structure of the developed material.

2. Experimental Details

2.1. Source of Materials

HCFC slag, the main ingredient for Geopolymer, is collected from **Shyam ferroalloys** (Durgapur, West Bengal, India). Sodium hydroxide (NaOH) is used as an activator, which is obtained from M/S Loba Chemicals, India. Sodium silicate (Na_2SiO_3) is used for the purpose of polymerization and is procured from Merck, India. Water soluble plasticizer (Sika) is purchased from a reputed firm located in the southern parts of India (Visakhapatnam market). Plasticizer usually enhances plasticity of the product *i.e.*, Geopolymer.

2.2. HCFC slag based Geopolymer (GP) preparation

Geopolymer is prepared from HCFC slag used sodium silicate, and alkali as additives. HCFC slag is crushed and ground to 240 mesh size. The fine particles are then dried in an oven for 2 h at 120 °C. The ground and dried materials are mixed with requisite amount of NaOH (8 M), Na_2SiO_3 and water soluble plasticizer (Sika). The semi-solid liquid materials, thus formed, are placed in a mould which is cylindrical in shape (55 mm diameter and 70 mm height). The mixture is rammed in the mould, then uniformity of which is ensured by continuous ramming the mass in the mould three times. Following ramming of the mass in the mould, the samples are left for 3 min for setting before demoulding the green sample [11]. The entire process is shown in

Fig. 1. The green samples are then cured at different temperature *i.e.,* 60, 70, 80 °C for different length 7 days and 28 days. Prepared samples are then characterized by XRD, FTIR TGA, DSC, etc.

Figure 1 Schematic diagram of process for preparation HCFC slag based GP product

2.3. Characterization techniques.

X-rays diffraction studies are made with treated GP samples at different stages of curing. Similar analysis is also made with the as received HCFC slag for comparison purpose. 1710 advance wide angle XRD is used with CuKα radiation for all XRD diffraction studies.

Samples are further characterized in **FTIR** (Thermo Nicolt Nexus 870 spectrophotometer). The settings for the instruments are kept constant (50 scan at 4 cm^{-1} resolution, absorbance mode) during entire examination.

For SEM studies of virgin HCFC sample as well as the cured HCFC samples are coated with gold before examining in a scanning electron microscopy (**SEM,** Carl Zeiss Supra 40). The elemental analyses are carried out with the help of EDX attached to the microscope.

The uncured and cured materials are also run in a thermogravimetric (TG) analyser Perkin Elmer Pyris Diamond analyzer. Heating rate is kept 10 °C/min and experiments are performed in an inert atmosphere (N_2 gas). The entire TGA experiment s are carried out in the temperature range of 50-300 °C. For comparison purpose, TG run is also taken from as-received HCFC slag.

Differential scanning calorimetric (**DSC**, Perkin Elmer Pyris Diamond) study is made for measurement of enthalpy of the two materials *i.e.*, as-received HCFC slag and GP prepared with HCFC slag. The parameters such as heating rate (10 °C/min), run temperature (50 to 300 °C), and nitrogen environment (N_2) are set while run is taken.

Compressive strength of prepared HCFC slag based GP is determined following ASTM standard in a tensile testing machine (**AIMIL COMPTEST** 2000, India). Maximum load is taken at the point of fracture of the sample.

3. Results and Discussion

The **Table 1** shows constituents such as SiO_2, Al_2O_3, Fe_2O_3, and Cr_2O_3, which are present in major quantities and CaO, Cr_2O_3, FeO in minor quantities [10]. The ground slag (240 meshes) is dried for 2 h at 120 °C to make it free from moisture.

IOP Conf. Series: Materials Science and Engineering **410** (2018) 012002 doi:10.1088/1757-899X/410/1/012002

Table 1 Chemical composition of HCFC slag

Constituents of HCFC slag with percentages	SiO_2 =28-31 MgO=26.0-26.5 Al_2O_3=23.5-24.5 CaO =7.5-8.5 Cr_2O_3=4.5-5.4 FeO= 2.5-3.0
Sources	Syam Ferroallys, Durgapur

Effect of molar concentration on compressive (MPa) study is shown in **Figure 2**. Subscribes 1, 2, 3, and 4 attached to S in the **Fig. 2** represent Geopolymer prepared with different molar concentration of NaOH *i.e.,* 6, 8, 10, and 12 which are cured at 70 °C. It is well known that sodium plays an important role for polymerization of alumina (Al_2O_3) and silica (SiO_2) present in the slag [4-8].

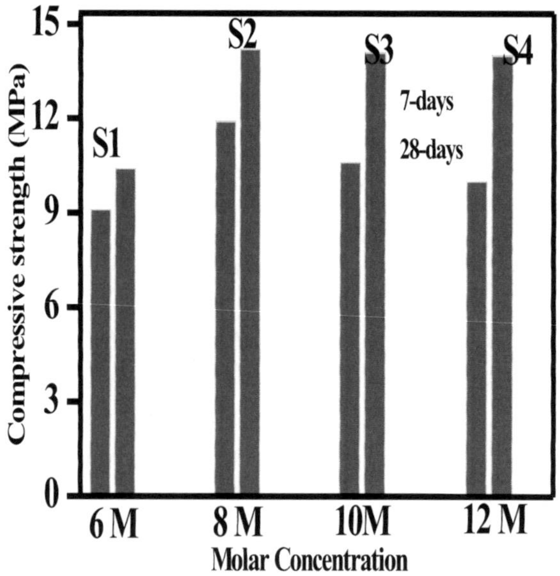

Figure 2 Compressive strength vs molar concentration at 7-days and 28-days curing

IOP Conf. Series: Materials Science and Engineering **410** (2018) 012002 doi:10.1088/1757-899X/410/1/012002

Each pair of bar drawn is showing data on comparative scale for a Geopolymer prepared with some molar concentration and cured for 7 days and 28 days. It is evident that maximum strength is obtained for the Geopolymer formed with 8 M NaOH concentration.

Similar bar diagram is shown in **Figure 3**. Here subscribes attached T designated Geopolymer prepared with 8 M NaOH and cured for different temperature for two curing time *i.e.*, 7 days and 28 days. The maximum strength is obtained for 15 MPa cured at 70 °C for 28 days [12].

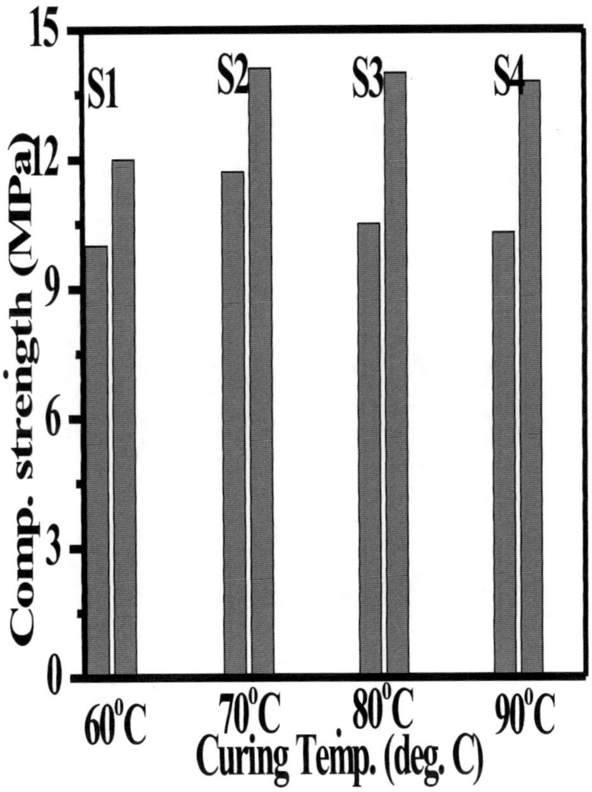

Figure 3 Compressive strength vs temperature at 7-days and 28-days curing

Thus, optimum combination of cured parameter obtained is curing 70 °C for 24 h and with NaOH of 8 M concentration.

Fig. 4 shows superimposed XRD pattern obtained from three different samples, designated 4A and 4B. **XRD** for as-received HCFC slag (**Fig.4A)** shows presence of sharp peaks, indicating crystalline phases, identified as quartz and mullite. Sharp peaks without line boarding essentially show crystallinity of those phases. XRD pattern for GP after curing at 70 °C is shown in **Figs.4B**. Sharp crystalline peaks have disappeared after curing at 70 °C. The disappearance of crystalline phases and appearance of non-crystalline phase are striking features of the diagram. There are broadened line profiles occurring at different Bragg's angles. This is co-related with polymerization process of slag, during curing.

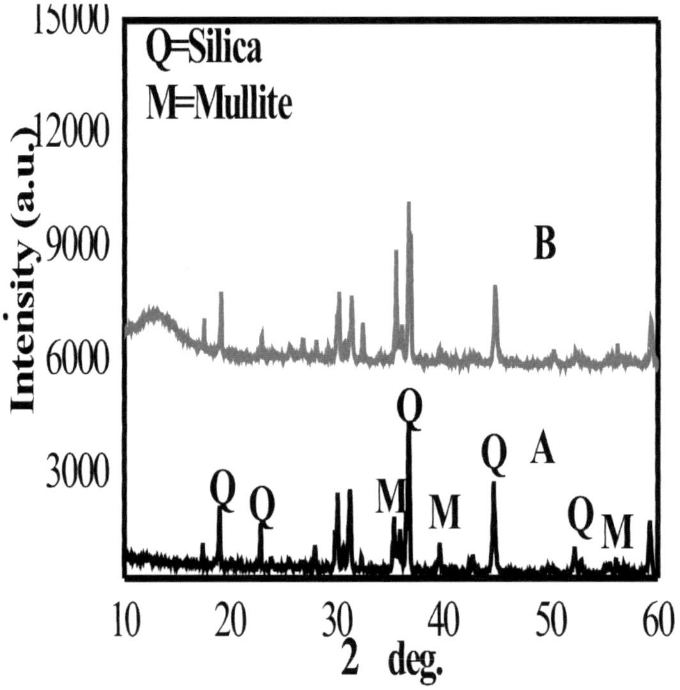

Figure 4 XRD pattern of as received HCFC slag (A) and as-prepared HCFC slag GP

The **microstructural features** of HCFC slag and corresponding slag based GP samples are observed under Scanning Electron Microscopy (SEM). The electron micrograph of unreacted slag and cured HCFC slag based GP are shown in **Figure 5**. **Fig. 5(A)** shows SEM micrograph of HCFC slag.

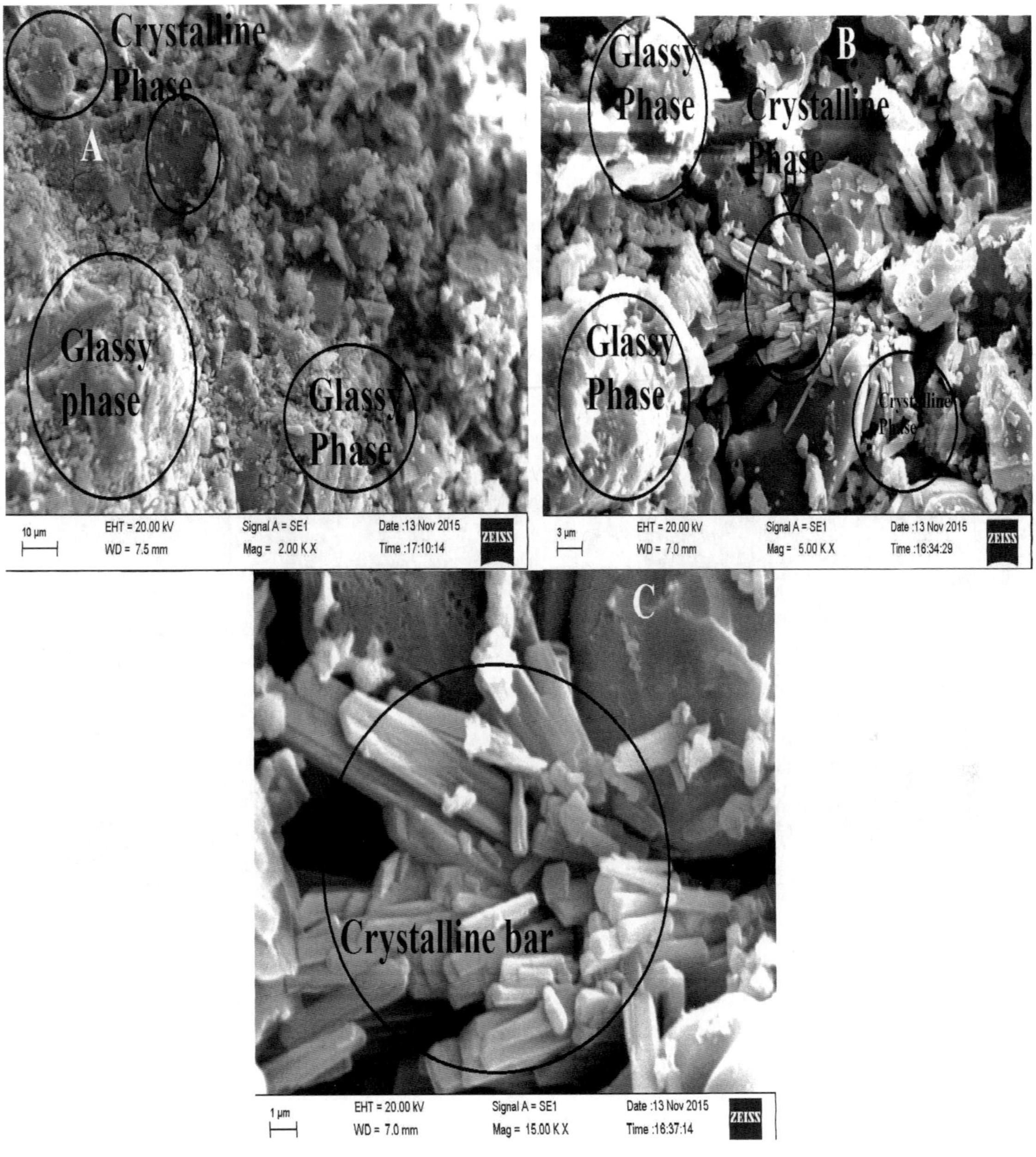

Figure 5 SEM images of as-received HCFC slag (A), Cured HCFC slag based GP

It indicates fused mass of oxide with some glassy phase spread over the matrix. Since the materials are obtained from stainless steel industries, therefore different minerals are fused and

IOP Conf. Series: Materials Science and Engineering **410** (2018) 012002 doi:10.1088/1757-899X/410/1/012002

are appearing in the microstructure. There is a someplace marked is as existing glassy phase owing to faster cooling of molten slag to room temperature.

Fig. 5(B) shows micrograph of Geopolymer prepared by curing of a green samples. The material after treatment with alkali materials cured at 70 ºC and 24 h have yielded different microstructure. At lower magnification, it shows some crystalline-needle shape phase in random direction and marked. There is some region of glassy phase co-existing with the crystalline phase. The **Fig. 5(B)** looks very interesting, therefore further investigated as different location at higher magnification.

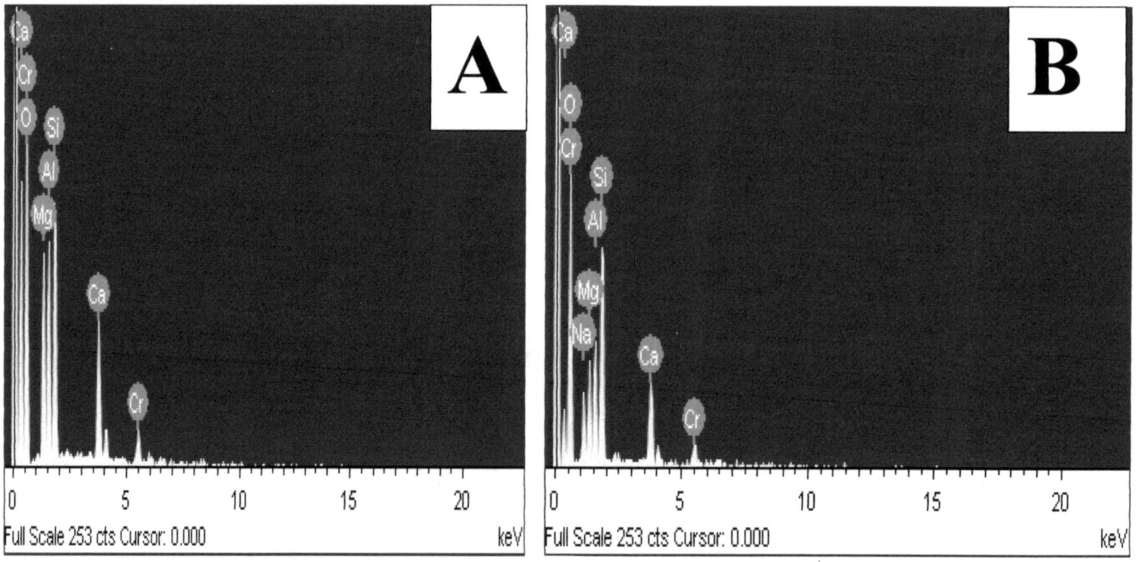

Figure 6 EDS of as received HCFC slag (A) and Cured HCFC slag based GP (B)

At higher magnification, there are bundles of rod-shaped crystalline phase orienting in same direction proving or the polymerization process [13]. Similar structure is also observed by S.D. Muduli et al. [13]. This crystalline phase occurred due to crosslinking of alumino-silicate Geopolymer phase and the process is assisted by sodium. The clearly glassy phases are visible adjacent to crystalline phase.

The in-situ EDS analysis is done on crystalline and glass phases and is shown in Figure 6. The presence elements such as Na, O, Al, Ca, Mg, Cr, etc are present both in crystalline and glassy phases of cured HCFC slag based GP (B), whereas O, Al, Ca, Mg, Cr, etc are present in as-received HCFC slag. Presence of Na and intensity ratio of element present signifies the chemical reaction.

FTIR spectrum of HCFC slag (**Fig. 7A**) and as-prepared GP (**Fig. 7B**) are shown in **Fig. 7**. The spectra indicate the presence of different bands appeared in each category of materials. For HCFC slag, bands occur at different wave numbers *i.e.*, 3441, 2918, 1638, 1442, and 887 cm^{-1}.

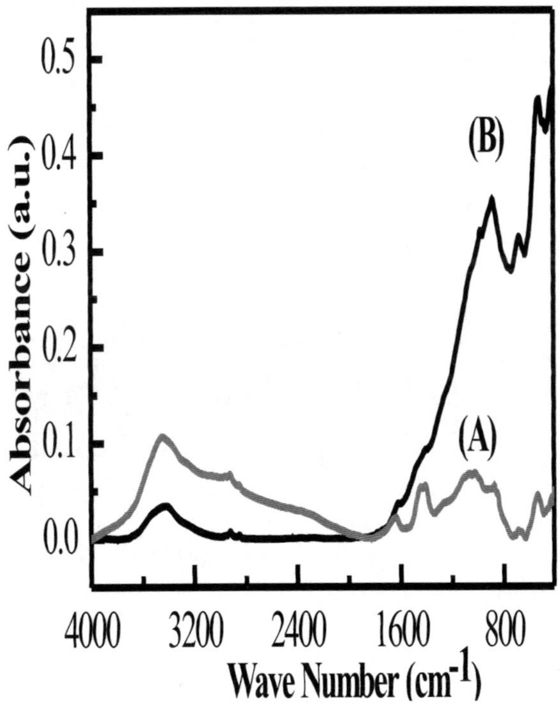

Figure 7 FTIR spectra of as-received HCFC slag (A) and Cured HCFC slag based GP (B)

Bands corresponding to 3441 and 1638 wave number are attributed to the stretching vibrations of O-H bonds and H-O-H bending vibrations of interlayer adsorbed H_2O molecule,

IOP Conf. Series: Materials Science and Engineering **410** (2018) 012002 doi:10.1088/1757-899X/410/1/012002

respectively [14]. The water hydroxyl-stretching plays an important role and so the intensity and peak shift of the infrared spectra is significant. Bands occur at 887 cm^{-1} is attributed Si-O band and it signifies the presence of silicate groups. Presences of $Al^{3+}O^{2-}$ bonds are also observed near 805 cm^{-1} [14]. Stretching vibrations is found at 440 cm^{-1} signifies the presence of Fe-O bonds [15]. Similarly, the peaks are not found for the as-received samples. This is attributed to the formation of Geopolymer due to treating and curing of as-received samples.

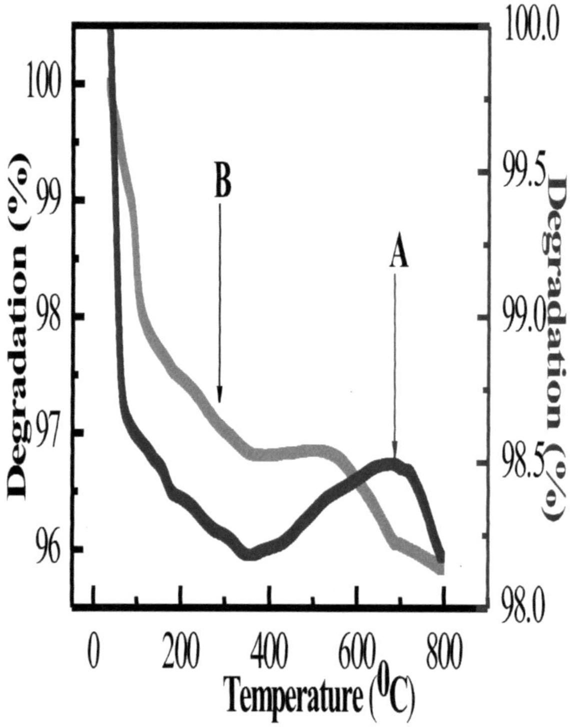

Figure 8 TGA curve of as-received HCFC slag (A) and Cured HCFC slag based GP (B)

In the TGA experiment, the weight loss was measured with increasing temperatures from 50-800 C and is shown in **Fig 8**. Looking to the above results **TGA run** is taken from two samples *i.e.*, as-received materials and cured GP materials striking difference can be observed. From the Fig. 8, Pure HCFC slag shows two types of weight loss. The sharp decrease in weight after 100 °C as recorded in the TGA thermograms, which is attributed to the loss of evaporable water presence in the form of moisture in pure HCFC slag. After the rapid decline of pure HCFC slag, the rate of weight loss stabilized between 100 °C-350 °C. Little change occurred in the percentage of weight remaining beyond 350 °C. The average total percentage of weight increase and decrease within the range of temperature 350 °C-700 °C exposures was 96% [16].

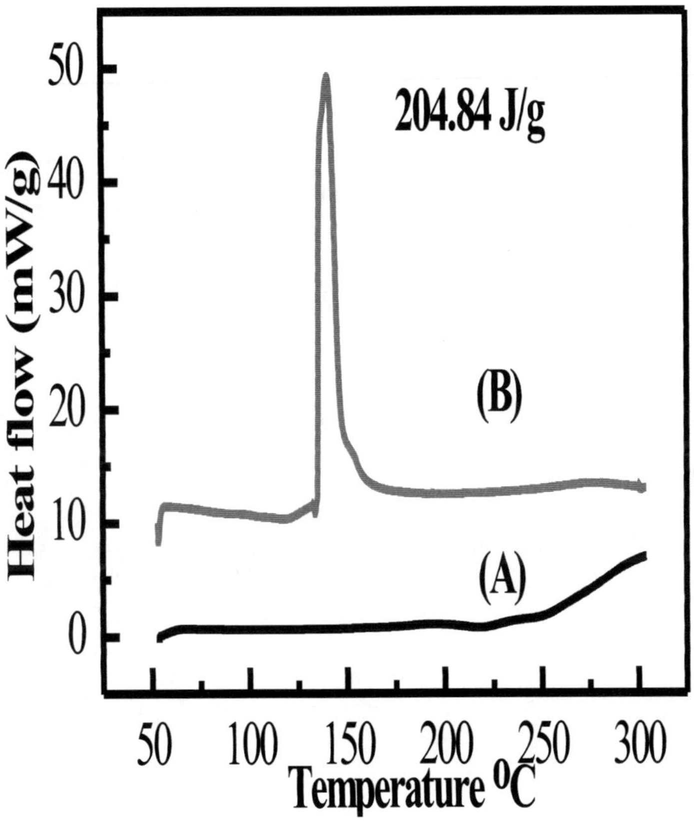

Figure 9 DSC isotherm of as-received HCFC slag (A) and Cured HCFC slag based GP

Figure 9 shows DSC isotherms for untreated and treated HCFC slag. It can observed that there no change of heat flow for the untreated slag (**Fig. 9A**), whereas there is a significant change in heat flow curve at around 141 °C in treated one. This is due to expulsion of physically bound water within the materials [17].

4. Conclusions

Geopolymers are developed from HCFC slag by treating with sodium silicate, sodium hydroxide, and water soluble plasticizers. The microstructures reveal formation of new glassy phase as reaction product with crystalline phase in between. XRD patterns shows decrease in the intensities of the phases as identified for the as-received samples. In a few phases, some peaks are conspicuously absent. Some line-broadening curves have seen indicating formation of glassy phase in place of crystalline phase. There is a marked difference observed in the FTIR spectrums. The number of peaks of as-prepared samples are much more comparison with as-received samples. This is due to formation of new bonds within the phases. TGA curves have revealed that both materials there are a gradual degradation with temperature. However, Geopolymer materials are more stable as compared to the as-received samples. DSC results shows that there is a exothermic peak in Geopolymer at 150 °C which results of elimination of water form during condensation polymerization process. The compressive strength is maximum curing at 70 °C and 8 M sodium hydroxide keeping other variables are constant.

Acknowledgements

The Authors gratefully acknowledge the **GIET Gunupur** for providing the facility and financial support to carry out the works. The authors are also acknowledging Dr Pradeep Kumar Rana for their support.

References

1. A. Bilodeau, and V. M. Malhotra, 2003, *American Concrete Institute Materials Journal* Vol. 1, pp. 41-48.
2. L. Krishnan, S. Karthikeyan, S. Nathiya, and K. Suganya, 2014, *International Journal Research in Engineering and Technology* (IJRET), Vol. 3, pp. 2321-7308.
3. A. Allahverdi and E.N. Kani, 2009, *International Journal Civil Engineering*, Vol. 7, pp. 154-160.

4. M. Al Bakri, A.M. Kamarudin, H. Bnhussain, M. Nizar, and I.K. Mastura, 2011, Asian J. Scientific Research, January.
5. Z. Zhang, J.L. Provis, A. Reid, and H. Wang, 2014, *Constructional and Building Materials*, Vol. *56, pp.* 113-127.
6. M.I. Khan, K. Azizli, S. Sufian, A.S. Khan, H. Ullah, and Z. Man, 2014, *World Sustainability Forum–* Conf. Proceedings Paper.
7. V. Kumar, and V.K. Singh, and A. Srivastava, 2013, *Journal of American Ceramic Society,* Vol. *96,* pp. 2124-2131.
8. P.H. Kumar, A. Srivastava, V. Kumar, M.R. Majhi, and V.K. Singh, 2014, *Journal of Asian Ceramic Society,* Vol. *2*, pp. 169-175.
9. P.H. Kumar, A. Srivastava, V. Kumar, and V.K. Singh, 2014, *Journal of Asian Ceramic Society*, Vol. *2,* pp. 371-379.
10. M. Erdem, H.S. Altundogan, M.D. Turan, and F. Tumen, 2005, *Journal of Hazardous Materials B*, Vol. 126, pp. 176-182.
11. M. Panigrahi, P.K. Rana, S. Gupta, M.B. Kumar, and R.R. Dash, Proceedings of 19th Int. Conf. on Non-ferrous Met.-2015, Eds. B.K.Satpathy, Abhilash, I.N.Bhattacharya, pp 201-210.
12. G. Naidu, P.B.J. Kumar, S. Adisesh, and P.V.V. Satyanarayana, 2012, *International Journal of Computational Engineering Research,* Vol. *2*, pp. 912-918.
13. S.D. Muduli, J.K. Sarangi, B.D. Nayak, and B.K. Mishra, 2013, *Greener Journal of Physical Science,* Vol. 3, pp. 204-211.
14. J.D. Russell, 1987, Infrared methods. In: A hand book of determinative methods in clay mineralogy. *Wilson MJ (Ed.) Blackie* and Son Ltd, NY, Vol. 133, pp. 11-67.
15. K. Ramaswamy, and R. Venkatachalapathy, 1992, *Indian Journal of Pure Applied Phys*ics, Vol. *30*, pp. 171-xxx.
16. A.M.M.A. Bakri, H. Kamarudin, M. Bnhussain, I.K. Nizar, A.R. Rafiza, and Y. Zarina, 2012, *Review on Advanced Materials Sci*ence, Vol. 30, pp. 90-97.
17. P. Chindaprasirt, C. Jaturapitakkul, W. Chalee, and U. Rattanasak, 2009, *Waste Management,* Vol. *29*, pp. 539-543.

Finite element analysis of the dental crown: a case study of alumina based incisor

Pratik Samal[1], Sushree Swati Mohanty[2], Muktikanta Panigrahi[3], Saralasrita Mohanty[*,4]

[1] School of Mechanical Engineering, Vellore Institute of Technology, University Vellore-632014

[2] School of Life Science, Ravenshaw University, Cuttack-753003

[3] Metallurgical Engineering Department, Gandhi Institute of Engineering and Technology (GIET), Gunupur, Rayagada, Odisha-765022

[4]School of Physical Sciences, National Institute of Science Education and Research, HBNI, Jatni-752050, India

Abstract. The purpose of this study was to evaluate the stress distribution on newly developed alumina based dental crown (incisor) for understanding its biomechanical effects by Finite Element Analysis (FEA). Alumina based dental crown was fabricated through green machining of machinable alumina cylindrical compacts by Computer Numerical Control (CNC) followed by sintering at 1550°C. Alumina crown (Incisor) was scanned by using 3D laser scanner and modified subsequently before assembled in a 3D finite element model. The 3D model was constructed and analyzed under ANSYS environment. The model developed high stress values at the upper frontal and basal region. The large stress concentration factors and narrow profile of the upper region and the arrested base are believed to have been responsible for the high stress values. Since the stresses were lower than the compressive strength of the material, the crown is expected to withstand real-time static loading.

Keywords: Alumina, Dental crown, FEA, CNC

[*]Corresponding author, email-saralasrita@niser.ac.in

IOP Conf. Series: Materials Science and Engineering **410** (2018) 012003 doi:10.1088/1757-899X/410/1/012003

1. Introduction

The property of tooth mostly depends on the sign of load, loading rate, and the direction of the applied load with reference to the microstructure orientation. The effect of contact stress on the dental crown associated with the oral functions, influences the lifetime of dental restorations. Currently, study of stress distribution pattern in dental implants or crowns has become a topic of great interest due to steep rise in tooth replacements. Stress analysis of fabricated dental structures provides information related to biomechanics and skeletal anchorage system to the dentist. Further, it helps in analyzing the physical, chemical and biomedical behavior of materials being used in implant dentistry that eventually leads to predict the longevity of the dental implant. Wide range of methods have been explored for stress measurement and analysis, including strain gauge method, loading test, photo-elastic method and finite element method. However, the conventionally used methods such as strain gauge, photo-elastic methods and loading test are not feasible as they are time consuming and expensive. It is also difficult to obtain analytical solutions for the dental crown owing to the sheer complexity of dental structures. In this context, recently Finite Element Analysis (FEA) is widely used for mechanical analysis of dental implants/crown. Use of FEA though not entirely accurate, it overcomes most of the shortcomings of the conventional techniques. FEA has been widely used by contemporary researchers to study the behavior of crowns and implants in static [1] as well as dynamic [2] scenarios. FEA of crowns made of glass has been carried out by Hojjatie and Anusavice using a three dimensional model [3]. Ambulgekar and Parle have carried an FEA on a premolar tooth to study its stress pattern [1]. Jager et al. have used FEA to study the effect of varying design parameters and core materials for a ceramic dental crown [4]. Li et al. have used a finite element model to assess the reliability of ceramic dental crowns [5]. From their studies it can be concluded that FEA is a valid method to analyze complex geometries and to study dental stress distributions as it overcomes several limitations of traditional techniques. FEA can also be used to study the mechanical behavior of implants/crowns made of recently developed materials.

It is to be noted that in 2013 Mohanty et al. successfully managed to fabricate a dental crown via a novel green state CNC machining using diamond embedded tool [6]. The cylindrical alumina compacts were prepared using Protein Coagulation Casting technique. Though the material [7] used for the process has shown reasonable mechanical properties with respect to natural dental crown, however, analysis to determine the stress distribution for the crown itself has not yet carried out.

Thus, this study aims to develop the stress distribution patterns for the alumina based dental crown developed by Mohanty et al. through a Finite element model using ANSYS environment. For the analysis, two loading cases have been considered (vertical and horizontal) and the results were analyzed carefully. The magnitude of the force to be applied on the crown has been collected from a previous study [8].

2. Model Generation

Since the geometry of the incisor is not a simple one, trying to create it from scratch would have been time consuming and would have also lacked precision. Instead, an original cadaveric crown was scanned using the laser scanner. The images for the incisor were reverse engineered using reverse engineering software. The three dimensional image obtained was converted to .stl (Stereo Lithography) format which was used subsequently for the CNC machining.

Further, the prototype so manufactured was laser scanned again to obtain the model (in .stl format) used for this analysis. "Stl" file formats describe the surface geometry of the component whereas a volume/solid model was needed for the FE analysis. The stl file that was obtained was first imported to Solidworks for the generation of a solid model. The model was scaled down to the actual size of the prototype. For further work, the generated model was transferred to ANSYS Workbench as shown in figure 1 and figure 2.

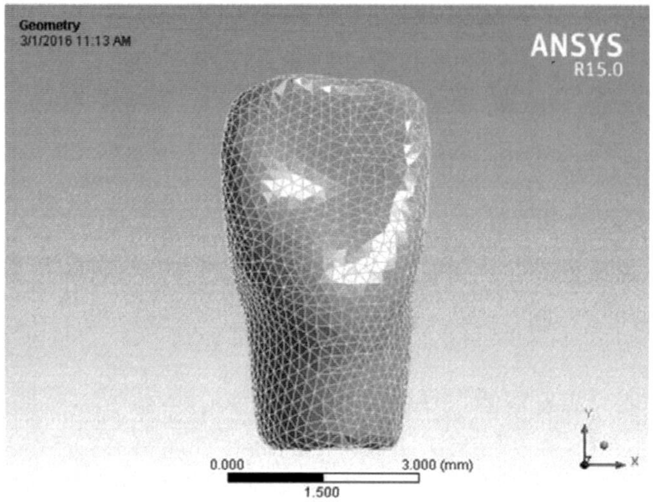

Figure 1. Front view of the dental crown in ANSYS Workbench

Figure 2. Side view of the dental crown in ANSYS WorkBench

3. Mechanical Properties of alumina

The compressive strength measurement of sintered aluimina was performed on five different cylindrical samples of diameter 13mm and 15mm in height, respectively by using **INSTRON 1344 (Capacity 500KN)**. The output obtained was, as follows.

Table 1. Mechanical properties of alumina obtained from the compressive test

Property	Value
Compressive Strength(MPa)	1037.71
Compressive stress at Yield (Offset 0.2 %) (MPa)	751.90
Young's Modulus (MPa)	23119.545
Compressive extension at Maximum Compressive load (mm)	2.69289
Compressive strain at Maximum Comp. load	17.95421 %
Poisson's Ratio	0.22

4. Meshing

In order to carry the analysis, it was first required to descretize the solid model into elements. The meshing was successfully carried out using ANSYS Workbench and the meshed model shown in figure 3(a) and figure 3(b). Element type of the mesh was SOLID 187. The node count of the meshed model was 130185, whereas the element count was 90934. The areas of the model which have irregular surfaces were selectively refined to get a better mesh quality without compromising the accuracy of the results due to increase in the mesh stiffness. A reasonable element quality of 0.83 was obtained. It must be mentioned that it was not possible to generate a perfect mesh for a geometry as complicated as that of a tooth. However, the parameters used were efficient enough to provide good results.

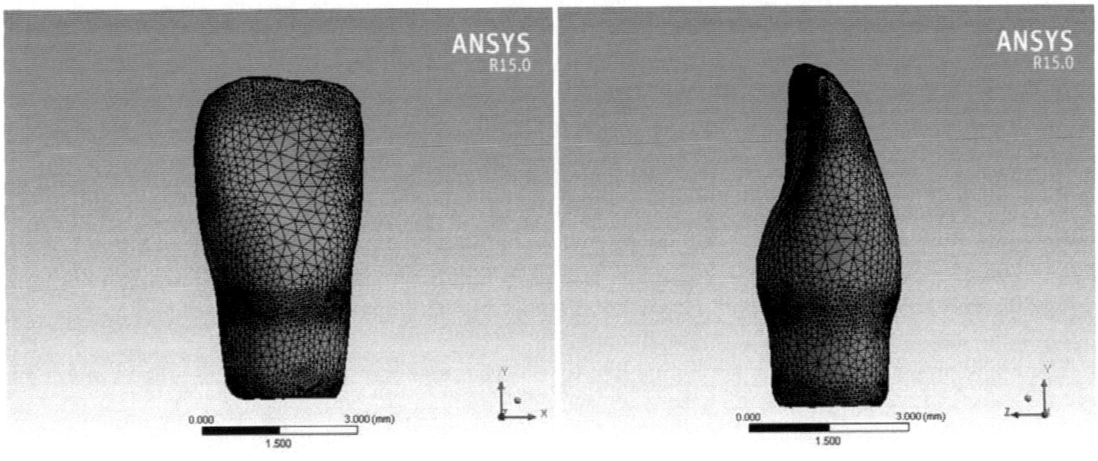

(a) (b)

Figure 3. (a) Front view and (b) Side view meshed model for the dental crown

5. Loads and Boundary condition

5.1. Application of boundary conditions

For applying the loads on the crown, following assumptions have been made. Human teeth have deep roots covered by gums, Hence for the analysis it was safe to assume a fixed support at the base i.e. to arrest all degrees of freedom for the base of the dental crown as indicated by the shaded blue region in figure 5 and figure 6.

5.2. Loading cases

The dental crown under study is that of an incisor. In a previous study to measure the biting forces the mean force for central incisors has been found out to be 189.3 N for normal teeth and 181 N after implantation [8]. Incisors are narrow edged teeth which are used for biting/cutting, and hence generally experience vertical loads. For the first case we have considered a vertical force 190 N in magnitude (figure 4). For the sake of completeness another case for the application of a shear force has also been considered albeit with a lesser magnitude (figure 5). The forces in their component form have been shown in table 2.

Table 2. Different loading conditions applied on the dental crown

Load Type	X component (N)	Y component (N)	Z component (N)
Vertical Force	0	-190	0
Horizontal Force	0	0	15

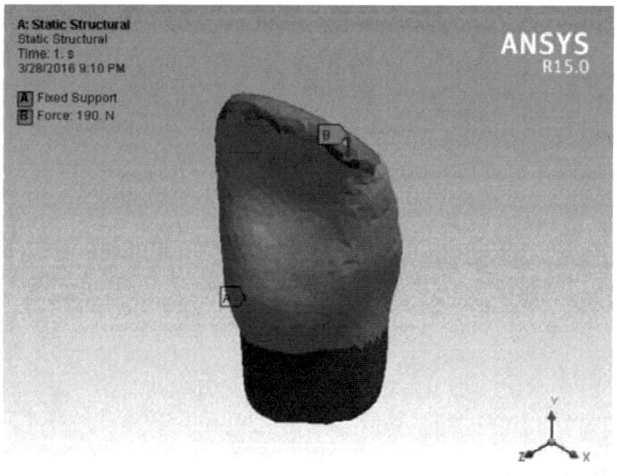

Figure 4: Application of vertical load for case-1 with fixed base

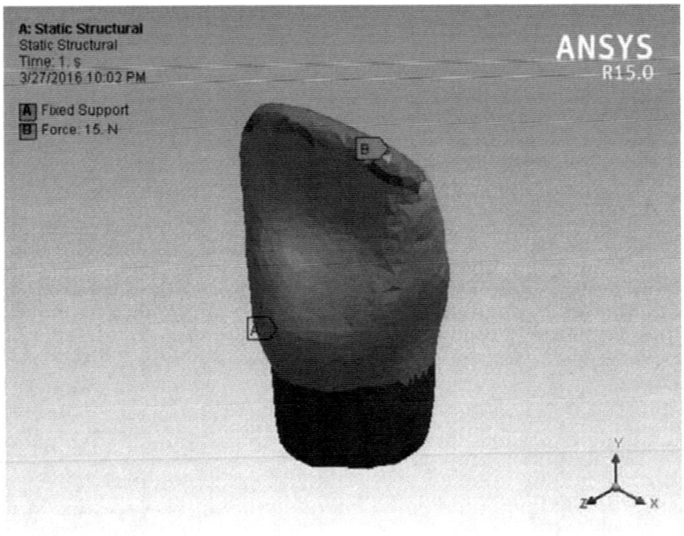

Figure 5: Application of horizontal force for case-2 with fixed base

6. Results and Discussion

The model was solved on ANSYS Workbench and the results for the Von-Mises Stress distribution were analyzed for both cases. The results have been shown in table 3.

Table 3. Maximum and minimum stress values for the two loading cases.

Load Case	Max Stress (MPa)	Min Stress (MPa)
Vertical Load	388.96	0.9198
Horizontal Load	149.92	0.0710

From figure 6 it can be clearly seen that the area just above the fixed basal region had the maximum stress values since under load, this region was unable to deform, because all degrees of freedom for the base had been arrested. But when the horizontal force was applied, the crown experienced bending about the base. As a result, in the second loading condition the frontal area was compressed, where as the rear portion was stretched which created a region of high stress near the base in the rear view as well (figure 6d). Additionally the entire front upper region of the crown under vertical loading was observed to be a region of high stress (figure 6a) for the following reasons. This region of the crown was highly irregular and had a large number of erratic contours which gave rise to large stress concentration factors. When viewed from the side we also notice that this portion was narrower compared to the base, which ultimately contributed to generation of stresses. This observation may be corroborated by a very common failure observed in dental crowns called 'chipping', wherein a portion of the tooth/crown breaks away from the main body. However the compressive strength of alumina and all brittle materials in general is very high. The stresses developed were very small compared to the compressive strength of the crown material (Table 1).

IOP Conf. Series: Materials Science and Engineering **410** (2018) 012003 doi:10.1088/1757-899X/410/1/012003

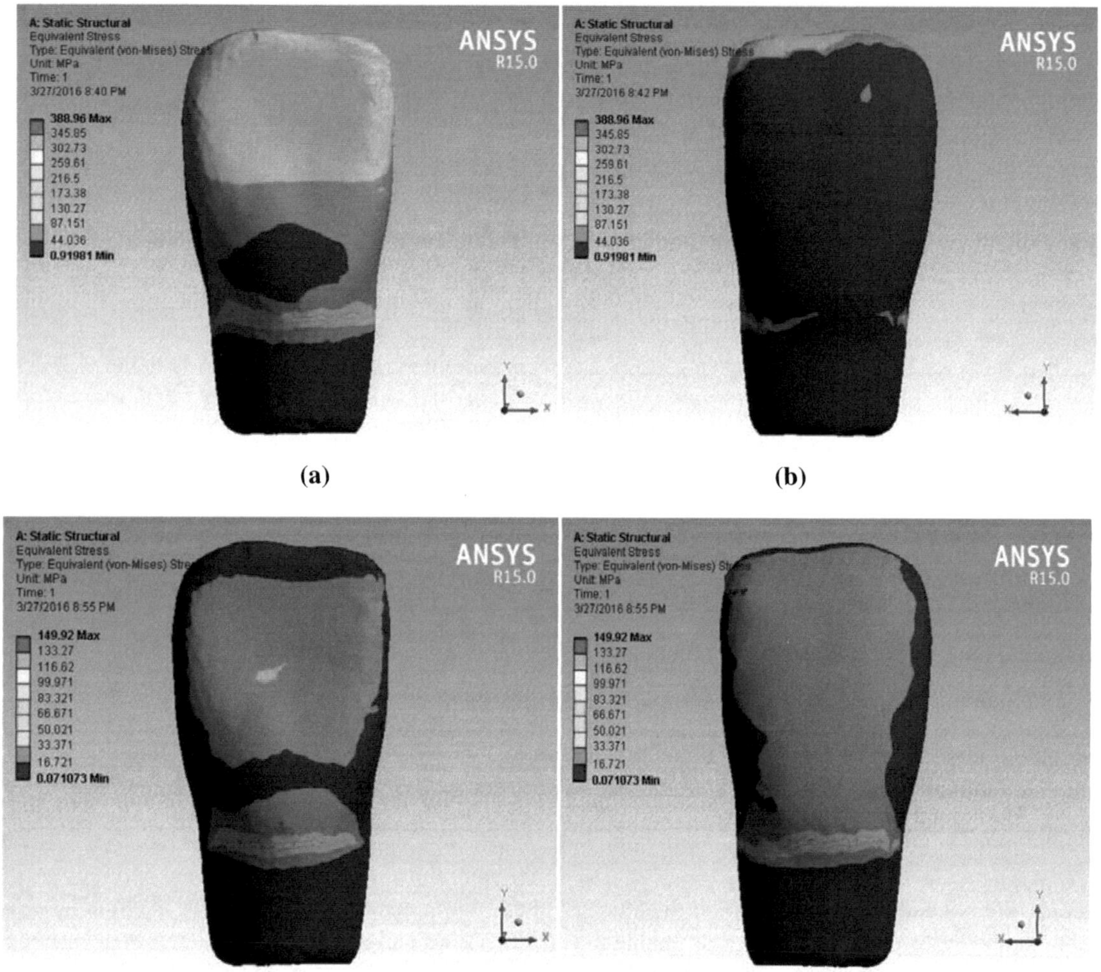

(c) **(d)**

Figure 6. Von-mises stress distribution for vertical forces (a) Front, (b) Back and Horizontal forces (c) Front, (d) Back, respectively.

IOP Conf. Series: Materials Science and Engineering **410** (2018) 012003 doi:10.1088/1757-899X/410/1/012003

7. Conclusion

The finite element analysis of the alumina based incisor dental crown was successfully carried out. From this study, the maximum equivalent von-mises stress developed for the dental crown under vertical loading (190 N) was found to be 388.96 MPa. Furthermore, when a horizontal load (15N) was applied alone, the model developed a maximum stress of 149.2 MPa. For both cases, the minimum stresses developed were very small, in the range of 0.05 to 1 MPa. It was also observed that the specimen's frontal and basal region were the sites of maximum stress concentration and potential sites of crack initiation. However, the maximum stresses generated for vertical and horizontal loading were 37.48% and 14.37% of the compressive strength of alumina (1037.71 MPa). Therefore, the dental crown was safe under static loading conditions. Hence, it was concluded that the alumina based dental crown developed by Mohanty et al. could be a feasible tooth replacement in implant dentistry.

Acknowledgements

The authors would like to thank Dr. Santanu Dhara, BMTE Lab, School of Medical Science and Technology, IIT Kharagpur-721302 for assistance with the fabrication and mechanical characterization of alumina based dental crown.

References:

[1] D. Parle, A. Ambulgekar, K. Gaikwad, , Infosys HTC, 2012

[2] D. Parle, D. Desai, A. Bansal, Estimation of Individual Bite Force during Normal Occlusion using FEA, Altair Tech Conf 2013, Infosys, India

[3] B. Hojjatie, and K.J. Anusavice, 1990, Journal of Biomechanics, Vol. 23. No. 11, pp. 1157–1166

[4] N.D. Jager , P. Pallav, and A.J. Feilzer, 2005, Materials Vol. 21, pp. 242–251.

[5] Y. Li, J. Chen, J. Liu, L. Zhang, W. Wang, and S. Zhang, 2013, Biology and Medicine, Vol. 43. No. 9, pp. 1214-1220.

[6] S. Mohanty, A. Rameshbabu, S. Mandal, B. Su, and S. Dhara, 2013, Journal of Asian Ceramic Societies, Vol. 1, pp. 274–281

[7] S. Mohanty, A. Rameshbabu, and S. Dhara, 2013, Ceramics International, Vol. 39, pp. 8985–8993

[8] B.K. Biswas, S. Bag, S. Pal, Biomechanical analysis of normal and implanted tooth using biting force measurement, IJEAS, August 2013. Vol. 4, No. 2

IOP Conf. Series: Materials Science and Engineering **410** (2018) 012004 doi:10.1088/1757-899X/410/1/012004

A Dielectric Barrier Discharge (DBD) Plasma Reactor: An Efficient Tool to Measure the Sustainability of Non-Thermal Plasmas through the Electrical Breakdown of Gases

S. P. Das*, G. Dalei and A. Barik

Department of Chemistry, Ravenshaw University, Cuttack-753003, Odisha

Email:dassmrutiprava@yahoo.in

Abstract

Gas discharges in plasma atmosphere are known to consist of a collection of different particles, mainly electrons, ions, neutral atoms and molecules. In this piece of research work we aim for the complete decomposition of volatile organic pollutants in NTP system. So the formation of plasmas in DBD systems and its sustainability are the main goal. The present need is to characterize the plasmas and optimization of the designed and fabricated plasma system under variable conditions. One of most important issue is the electrical breakdown of gases and to find the point of electrical breakdown. The volatile organic pollutants taken are Benzene, toluene, xylene, chlorobenzene, dichlorobenzene, nitrobenzene, methylene chloride etc. Self-sustainability of DBD-plasmas is explained for different VOCs under different experimental conditions taking helium and argon as carrier gases in terms of Paschen's curves. It explains the breakdown voltage as a function of the electrode spacing or gas gap (d), operating pressure (p), and gas composition [1]. The breakdown voltage is a function of the product of the pressure (p) and the inter-electrode distance (pd) also. It is verified that as the nature of substituent group changes, it varies the breakdown voltage and glow discharge zone following Townsend breakdown curve and eventually depends upon electron density of the system.

Keywords: Gas discharges, VOCs, DBD systems, Paschen's curves, Townsend breakdown curve, self-sustainable plasma.

1. INTRODUCTION

In 1889 Friedrich Paschen developed a law which is known as Paschen's law and the curve obtained from it is called the Paschen's curve, it explained the breakdown voltage as a function of the electrode spacing or gap (d), operating pressure (p), and gas composition [1]. The breakdown voltage is a function of the product of the pressure (p) and the inter-electrode distance (pd).

$$V_b = f(pd) \qquad (1)$$

V_b= is the breakdown voltage.

The mathematical formulation of Paschen's curve is derived from Townsend's assumption [2]. Description of the basic charge generation processes including electron impact ionization (α process) and secondary electron emission from the cathode due to primarily ion bombardment (the γ process), though other bombardment processes may play a vital role in this theory [3]. Historically, Paschen's curve has proved to be accurate for large gaps and at low pressures [4-5], but it is often acknowledged that it fails to explain the behaviour of curve at extremely low or high pd values [6-7].

1.1Conditions for self sustained discharge

Sustainability of plasma depends upon the applied voltage which must exceed the breakdown voltage of the gases. When this voltage is attained, the gases lose their dielectric properties and turn into conductors. It must satisfy the following relations:

IOP Conf. Series: Materials Science and Engineering **410** (2018) 012004 doi:10.1088/1757-899X/410/1/012004

$$1-\gamma\left(e^{\alpha d}-1\right)=0$$

$$e^{\alpha d}=\left(1+\frac{1}{\gamma}\right)$$

Paschen derived breakdown voltage from the above relationship as:

$$V_b=\frac{Bpd}{\left\{\ln\left(Apd-\ln\left[\ln\left(1+\frac{1}{\gamma}\right)\right]\right)\right\}} \qquad (2)$$

Finally expressing:

$$V_b=f(pd) \qquad (3)$$

A and B are the constants and can also be calculated as:

A $=\frac{\sigma_n}{kT_n}$, σ_n is electron-neutral collisional cross-section, T_n is temperature of neutral atom, k is the Boltzmann constant.

B = AV_i where V_i is the ionization potential and γ is secondary electron emission coefficient.

From equation (1) it is clear that breakdown voltage depends on the product of p and d for a given gas. From the Paschen curve (figure 1), it is understood that breakdown voltage is minimum for a certain *pd* product. But it is large for both small and large values of *pd.* This can be explained as there are too few collisions/ too large collisions at low and high values of *pd* respectively [8-19]. In order to optimize the newly designed DBD reactor, Paschen law and Townsend Theory were verified and the results were discussed in this piece of work.

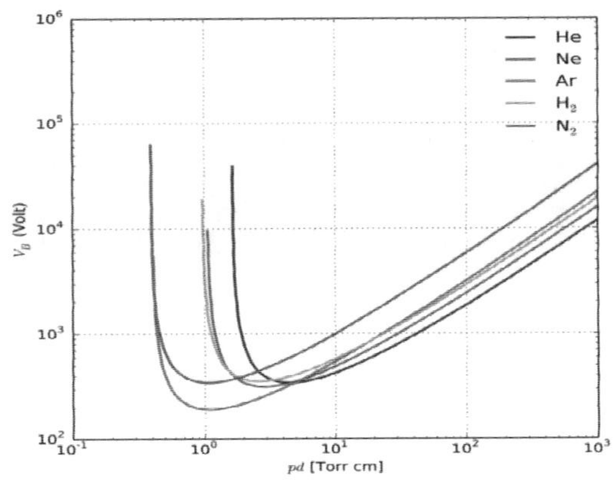

Figure 1. Paschen curves obtained for helium, neon, argon, hydrogen using the expression for the breakdown voltage as a function of the parameters A, B [20].

2. MATERIALS AND METHODS

In this work to optimize the DBD plasma system, we have performed a simple investigation on the electrical breakdown of Ar and He gas molecules between two parallel electrode plates separated by variable distance and voltage. The current is kept low so that the temperature is below the critical value beyond which thermionic emission occurs. In order to avoid emission of excess free electrons from the electrode, the low electric field is maintained. The parallel electrode plates are used to avoid corona discharge due to accumulation of charges on cornered surfaces that can cause high potential gradients around them.

The experimental arrangement used to study DBD discharge at atmospheric pressure condition is sketched in Figure 2.

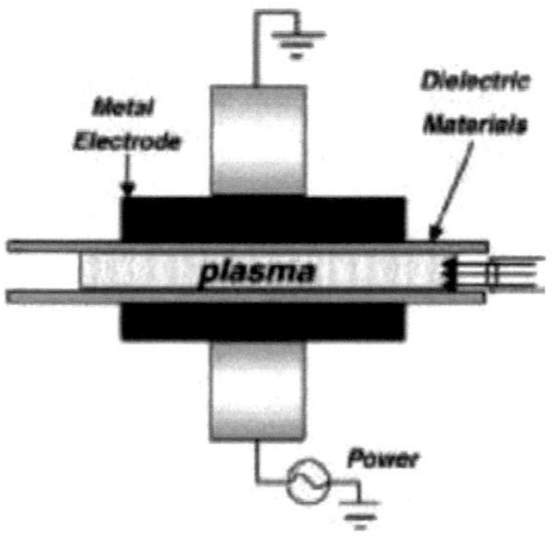

Figure 2 . Parallel Plate Plasma Reactor

The discharge is generated between two plane electrodes that is covered by 3mm thick pyrex dielectric plates. The normal AC power supply is connected (frequency 50Hz) to a step-up transformer for generation of potential from 0-30kV. Two mass flow controller of the range 0-5 lit/min is connected to measure the gas flow to the reactor. The breakdown voltage for different VOCs with carrier gases like He and Argon are studied here.

Two MFCs are connected to the carrier gases Argon and Helium. The other end is attached to the reactor chamber. When high voltage is applied to the system, plasma is generated. Varying different voltages as well as gas gap the system is optimized. The gas gap distance can be varied from 3mm to 15mm. preliminarily break down voltages at different gaps are noted. By plotting a graph between break down voltage verses pd (pressure (p) X distance (d)), Paschen Curves for different gases are obtained.

In the second case we changed the flow rate of the gases, and then measured the discharge current. Pressure exerted was calculated from the flow rate. Then by plotting graphs between break down voltages and (P X d), we get Paschen's Curve.

3. RESULTS AND DISCUSSION

3.1. *Optimization of the reactor in presence of Ar at a fixed gap.*

The optimization parameters for different gases (argon and He) varying voltage and distance were analyzed. Townsend breakdown and Paschen Curve were analyzed for different pressure and distance [6]. Impedance matching resistor was installed between HV power supply and DBD reactor for maximum power transfer at optimized operating condition.

The curve (figure 3) for Ar is obtained when gas-gap is kept fixed at 11mm and mass flow rate is varied from 0.5-5.0 lit/min. It is observed from the figure 4 that when P X d value is 24.75 (Torr X mm) the minimum value of V_b corresponds to 2.1 KV and at this condition the system is optimized.

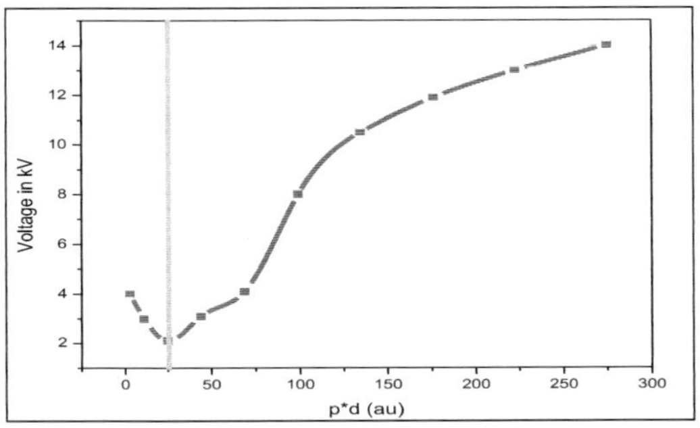

Figure 3. Paschen Curve for Ar

3.2 Optimization of the system in Ar with volatile organic compound.

In presence of Ar, different volatile organic compounds ionize in different voltage (figure 4, 5, and 6). Benzene, xylene and toluene and in Ar plasma ionize at 2.8 kV, 4.5 kV and 2.6 kV. Between benzene and toluene, due to branching, toluene reacts first. In spite of presence of two substituted methyl groups in xylene its breakdown voltage is higher than others. This may be explained in terms of steric effect. During the course of experiment we also found that xylene decomposes at higher voltages than benzene and toluene indicating presence of some stabilizing energy in xylene between methyl substituent.

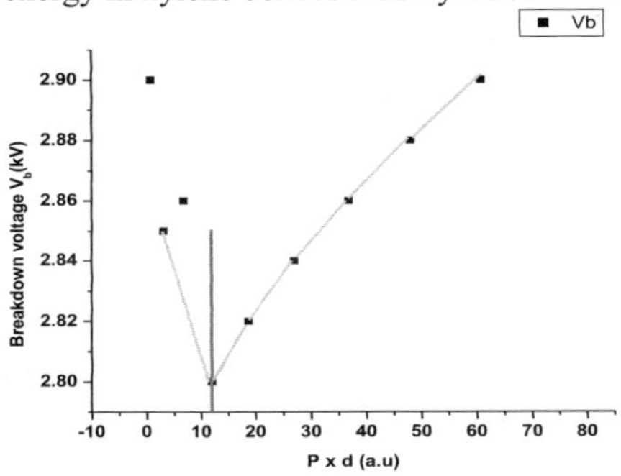

Figure 4. Paschen Curve for Ar + Benzene

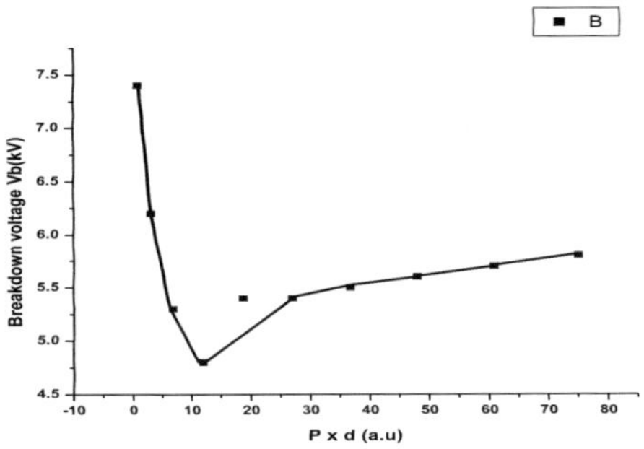

Figure 5. Paschen Curve for Ar + Xylene

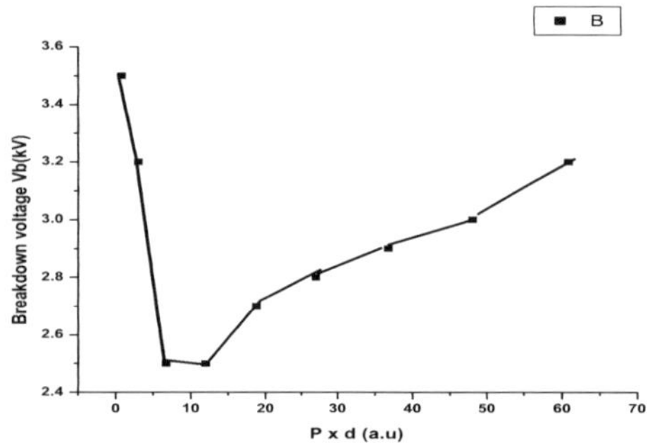

Figure 6. Paschen Curve for Ar + Toluene

3.3 Optimization of the plasma reactor depending upon the two different gap distances between the electrodes for He plasma

From the figures 7 and 8 it is clear that He plasma is best explained when its mass flow rate is 1.0lit/min. but the breakdown voltage is higher (6.5 KV) when the gas gap is 7mm (figure 7) as compared to 4mm distance (i.e 3.9 KV) (figure 8). The system can work efficiently for formation of sustainable helium plasma at a gas gap distance 4- 7mm and the applied voltage in the range of 3.9-6.5 kV. So the formation and existence of He plasma in this system can be understood in following ways:

i. It depends upon the distance between the electrodes.
ii. Also on the flow rate of the gas i.e. concentration of the gas present in the reactor volume
iii. Volume of the chamber as its volume depends upon the inter electrode distance

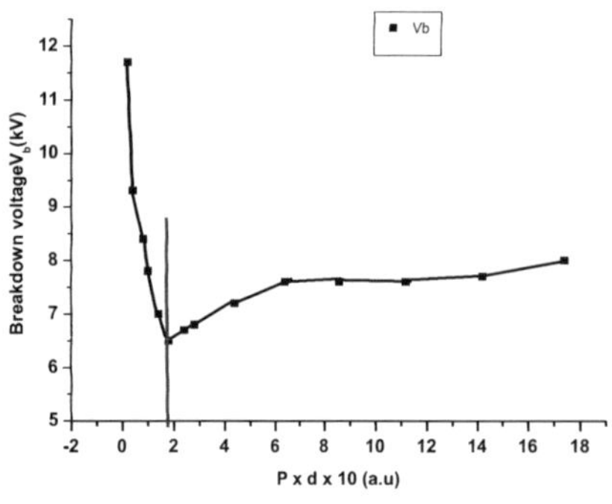

Figure 7. Paschen Curve for He (d=7mm fixed and mass flow is varied)

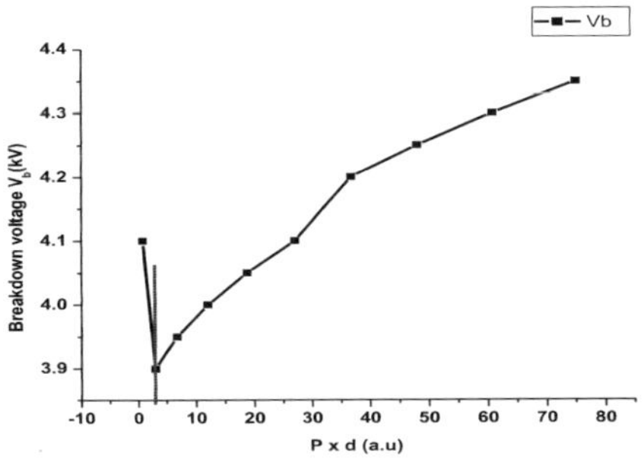

Figure 8. Paschen Curve for He (d=4mm fixed and mass flow is varied).

3.4 . Optimization of He plasma in presence of volatile organic compounds in different flow rates

From the figure 9, 10 and 11, it is clear that all the above system is optimized in the gas gap of 4mm for the formation of He plasma. In three cases mass flow rate is 1lit/min. The gas gap distance is 5mm and 4mm in figure 10 and 11 respectively. For figure 10 the breakdown voltage is 3.2kV and it is slightly greater than in figure-11 i.e 3.0 kV. The deviation may be due to the atmospheric humidity (condition) and gas gap. But in case of figure 9 as the gas gap (pd) is 8mm the minima broadened the Paschen curve and we get higher breakdown voltage at 5.5KV.

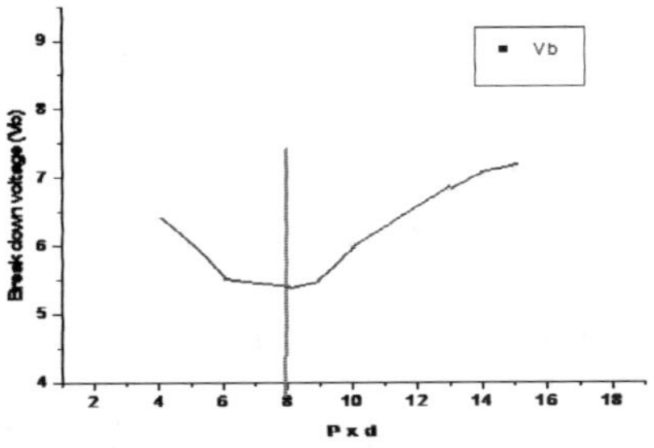

Figure 9. Paschen Curve for He + Benzene (d=8mm)

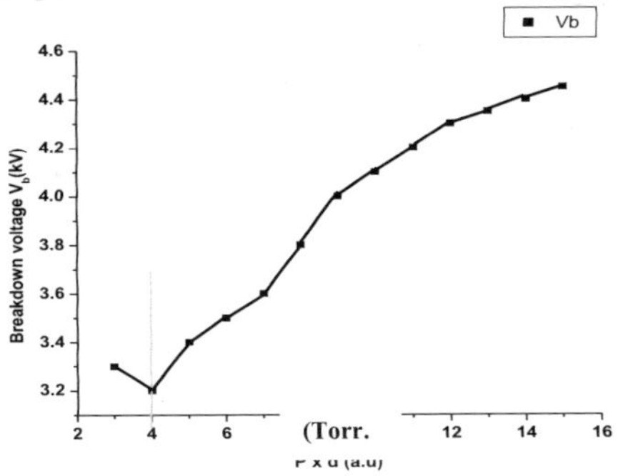

Figure 10. Paschen Curve for He + Benzene (MFC= 1lit/min) (d=5mm)

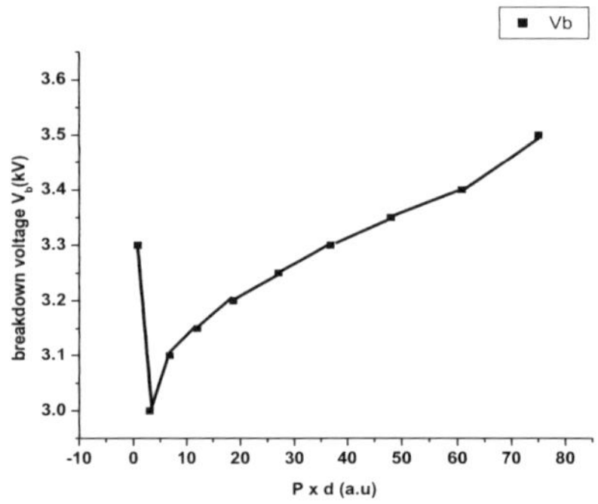

Figure 11. Paschen Curve for He + Benzene

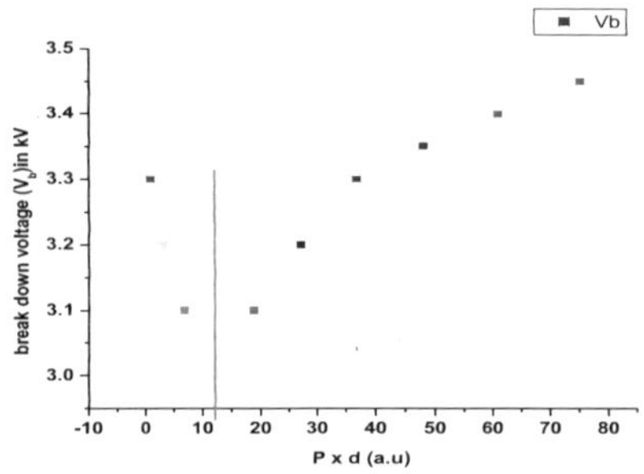

Figure 12. Paschen Curve for He + Toluene

3.5 Optimization of He plasma in reactor in the presence of different volatile compounds

Comparing the figures 11, 12, 13 and 14, the breakdown voltages of the volatile organic compounds like benzene, toluene, xylene and chlorobenzene are found to be 3.0kV, 3.0kV, 3.8kV and 2.7kV respectively. Which shows for a particular mass flow rate (1lit/min), the breakdown voltages differs for different compound. In case of nitrobenzene (in figure 15) the breakdown voltage is about 4.0kV which is much greater than chlorobenzene i.e.2.7kV. As the gap distance is more (6mm).So it required more voltage to break the compound at higher gas gap. It indicate breakdown voltage (i) depends upon the nature of the VOC compound and bond energies of the compounds.

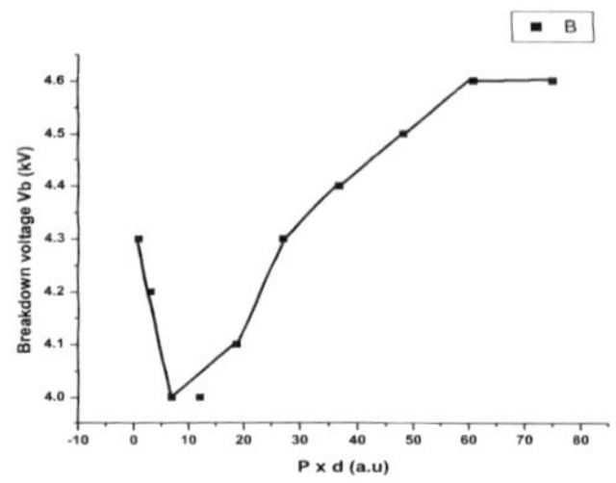

Figure13. Paschen Curve for He + Xylene

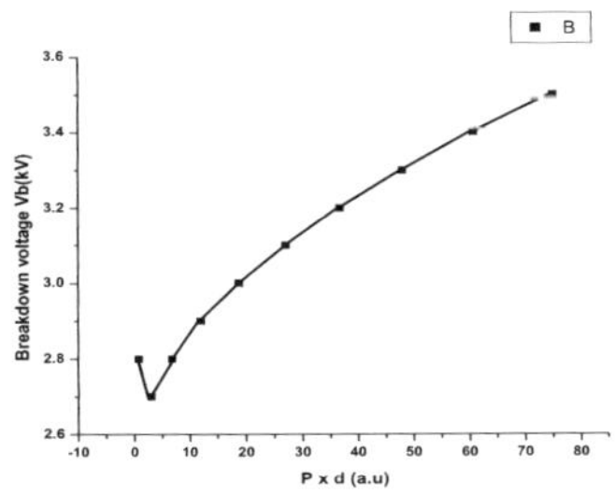

Figure 14. Paschen Curve for He + Chlorobenzene

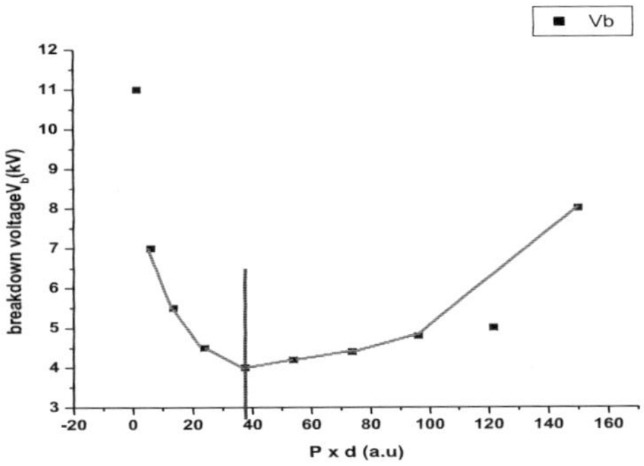

Figure 15. Paschen Curve for He + Nitrobenzene

3.6 . Verification of Townsend breakdown curve.

Plasma breakdown as an important fundamental process in plasma science has been a subject of enormous studies from the early days of gaseous electronics. Due to its relevance in a wide range of applications, it is used for a deeper understanding of fundamental plasma behaviour [21-24]. Renewed interest in breakdown phenomena, especially breakdown in small gaps, emerged from the possibility of lower facility and process costs for a variety of plasma processing and micro-manufacturing techniques. At the same time, direct current (DC), pulsed DC and radio-frequency (RF) discharges are widely used in the microelectronics industry, in plasma display panels, for depositing thin films, for semiconductor processing, surface modification, analytical chemistry, biotechnological and environmental applications, waste treatment etc. [25-31]. As already pointed out, a better understanding of voltage breakdown, besides being scientifically interesting, will aid progress in many fields and technologies, which generally fall into two categories: those that require high electric fields, and those that require high electric currents. On the other hand, unwanted voltage breakdown limits many technologies involving high electric fields [32]. Electric breakdown is referred to as a process that transforms a non-conducting material to a conducting one when a sufficient strong electric field is applied comprising an involved set of transient processes such as collision of electrons, ions and photons with gas molecules and electrode processes which take place at or near the electrode surface. Other possible gas processes include ion-atom collisions, excited atom-molecule collisions, and atom-atom collisions. In 1928, Langmuir [33] introduced the word plasma to describe the ionized gas that is created in a gas discharge.

IOP Conf. Series: Materials Science and Engineering **410** (2018) 012004 doi:10.1088/1757-899X/410/1/012004

From above facts here we conclude that
Mechanism: mechanism of Townsend theory involves three steps which is represented as below
Let A is a gas molecules or atoms undergo ionization involving following steps:

Step-1 photo ionization:

$$A + h\nu \rightarrow A^+ + e$$

Step-2 ionization by excited atoms or molecule

$$A + A^* \rightarrow A^+ + e + A$$

Or

Associative ionization

$$A + A^* \rightarrow A_2^+ + e$$

Step-3 Dissociative Recombination

$$A_2^+ + e \rightarrow A + A^*$$

Or **Associative recombination**

$$A^+ + e \rightarrow A + \gamma$$

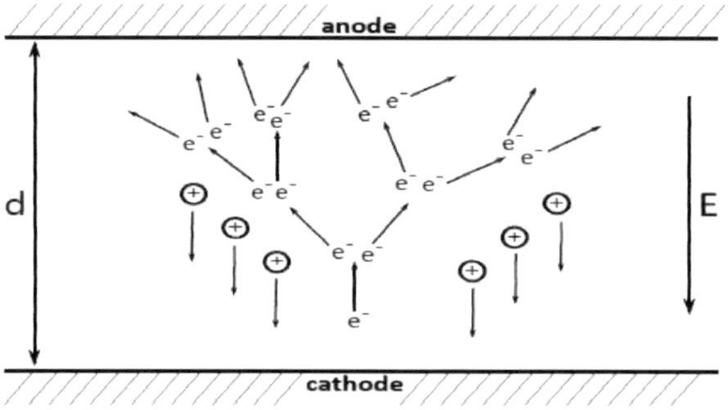

Figure 16. Electron avalanche on the basis of Townsend's theory

Following the Townsend theory (shown in figure 16) in this work, Voltage-current characteristic of low temperature discharge between electrodes for a wide range of currents is plotted in figure 17, plasmas become self-sustaining, i.e. the breakdown occurs. The simplest breakdown condition can be expressed as: each primary electron generated in the initial avalanche and lost at the anode is replaced by another electron generated by secondary emission at the cathode. This represents a steady self-sustained current in the homogeneous field:

$$Et = V_b/d$$

(Point B in Figure 17), where V_b is the breakdown voltage and is determined from Equation (3) as a function of d, γ and the known function α (EA: region of non- self-sustaining discharge, (BC) Townsend discharge, (CD) subnormal glow discharge, (DE) normal glow discharge EF) abnormal glow discharge. All these data shows the existence and sustainability of the plasma in the designed reactor.

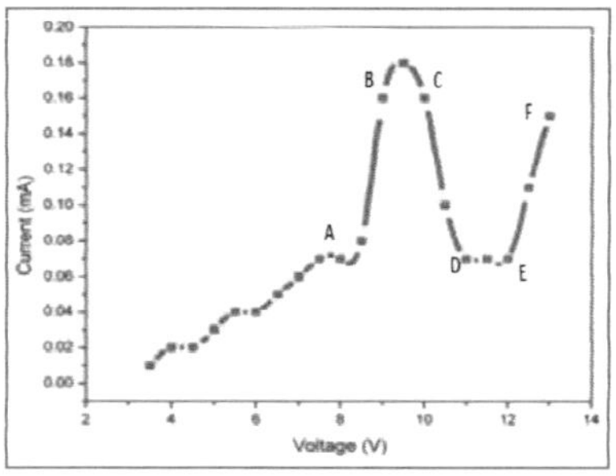

Figure 17. I-V curve of He + Benzene

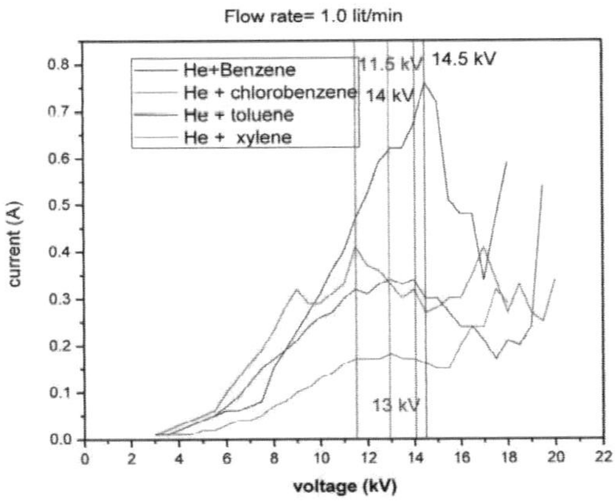

Figure 18. Breakdown voltage curve for different volatile compounds

3.7. Comparison of breakdown voltages for different substrates (VOCs) from current-voltage curve.

Influenced by Townsend experiment we have calculated the breakdown voltages (in figure 18 & 19). Breakdown voltage starts at 3.0kV, 2.7kV, 3.0kV and 3.8kV for benzene, chlorobenzene, toluene, xylene respectively in figure 18. But it is 3.2kV, 6.0kVfor ortho-dichlorobenzene and methylene chloride in figure 19. Here all these vocs are taken in combined with He as carrier gas. The bond dissociation energy of C-H bond in benzene is ~390kCal/mole [8], in toluene is ~77.5kCal/mol and in xylene ~77.5Kcal/mole [9], and that for C-Cl bond in chlorobenzene is ~399.19kCal/mole, for ortho- dichlorobenzene it is 385.4 kCal/mole and that for methylene chloride is 310.0 kCal/mole[10]. It is observed that C-Cl bond energy is highest, but its breakdown voltage is 2.7kV and 3.2kv in case of chloro and dichlorobenzene. For benzene bond energy is ~390kCal/mole, but dissociates at 3.0kV. Similarly, the glow region also starts at 11.5kV for chlorobenzene and 5.0kV for di-chlorobenzene where as that of benzene is 14.0kV. Since the electron affinity of Cl is appreciable, as soon as the chamber is filled with He plasma , Chlorine atom gets detached from chlorobenzene and dichiorobenzene quickly becoming neutral recombining with free electron in the plasma. Again for benzene the maximum current is drawn i.e. 0.67mA and for chlorobenzene it is lower than benzene i.e. 0.41mA whereas for dichlorobenzene it is much less i.e. 0.1mA justifying chlorine as a strong electron scavenger. In both the cases the chlorine ions get detached from chlorobenzene and dichlorobenzene in plasma medium by using DBD reactor at low breakdown voltage and current producing sustainable plasma.

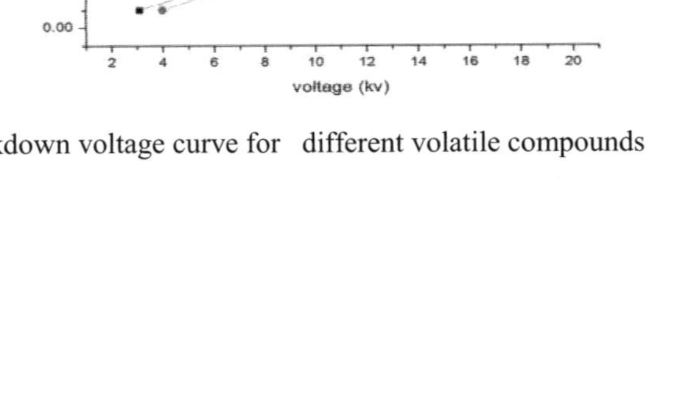

Figure 19. Breakdown voltage curve for different volatile compounds

But it is found that in case of aliphatic dichloro compound i.e. methylene chloride, bond dissociation energy is comparatively less i.e. 310.0kV whereas breakdown voltage starts at 6.0kV and glow region starts immediately at6.5kv. It may be due to low molecular mass, high volatile nature (B.P.-39.5) of the compound which gets dissociated completely in plasma medium due to electron induced reaction.

4. CONCLUSION

Dielectric barrier discharge assisted plasma technique is a novel green technique with high energy evolution. The developed DBD assisted plasma reactor is optimized using various diagnostic methods. Optimization of a plasma reactor depends on size, gap between the electrodes, gas flow rate as well as applied potentials and stability of carrier gases. It also depends upon the nature of the compounds and type of bond which has been verified taking different volatile organic compounds like benzene, toluene, nitrobenzene, chlorobenzene, dichlorobenzene, methylene chloride etc. Different aspects discussed above give the reliability of the Paschen curve and Townsend breakdown theory verified from our DBD plasma system operating in room temperature and atmospheric pressure.

ACKNOWLEDGMENT: Authors are thankful to BRNS India for funding and Ravenshaw University for providing all facilities.

REFERENCES

1. Paschen F. 1889 *Analytical Physics* **273** **5** 69-75

2. Townsend J. 1915 *electricity in Gases* (Oxford: Oxford University Press).

3. Townsend J. 1910 *The Theory of Ionization of Gases by Collision. Constable,*

   Online:http://www.worldcat.org/wcpa/oclc/8460026.

4. Dakin T.W Luxa G. Oppermann G. Vigreux J. Wind G. Winkelnkemper H. 1974. Electra. **32** 61-82.

5. Nasser E. 1971. *Fundamentals of Gaseous Ionization and Plasma Electronics (New York: Wiley-Interscience),*

6. Wadhwa C.L.2007 *High Voltage Engineering* (2nd ad.), New Age International, 10- 12. ISBN 8122418597.

7. Husain E. Nema R. 1982 *IEEE Trans. Electr. Insul.*, EI-17 **4**, 350-353,

8. Braithwaite N. St. J. 2000 *"Introduction to gas discharges". Plasma Sources Sci. Technol* **9** 517-527

9. Sen S. N 2006 *Plasma Physics Pragati Prakashan India* 8-12

10. Torres C Reyes P. G Castillo F. Martínez H 2012. *Journal of Physics: Conference Series* 370 012067

11. Davico G. E Bierbaum V. M. DePuy C. H. Ellison G. B Squires R. 1995 *J. Am. Chem. Soc.* **117** 2590-2599

12. Szwarc M 1948 *J. Chem. Phys*, **16** 2 128-136
13. Zhang R Li X Zhang X 2009 Chin. *J. Chem. Phys*, **22** 3 235-240
14. Hourdakis E. Simonds B. JZimmerman. N. M. 2006. *Rev. Sci. Instrum.*, **77** 3 034702-1-034702-4
15. Dash T.N Dey G R. 2015 BARC E 007
16. Dennison S Chapman A Lanagn M Hopwood J 2016 *Plasma sources Sci. Technol* **25** 03LT 02
17. Biswas J C Mitra V 1979 *Applied Physics* **19** 4377-381
18. .R. Tirumala, D. B. Go: *Applied Physics* Letters 97 (2010) 151502
19. M. Radmilovi´c-Radjenovi´c, B. Radjenovi´c: *Europhysics* Letters 83 (2008) 25001
20. Lieberman M. A. Lichtenberg A. J. 2005 *"Principles of plasma discharges and materials processing"* 2nd ed. 5-10.
21. Dario P. Carrozza M. C. Benvenuto A. Menciassi A. 2000. *J. Micromech. Microeng.* **10** 235-244
22. Ito T. Izaki T. Terashima K. 2001 . *Thin Solid Films* 386 300-304
23. Neumann A. 2001 *J. Appl. Phys.* **90** 1-26
24. Tsuyohito I Kazuo T , 2002 *Appl. Phys. Lett.* **80** 2648-2650
25. Pai Z D Stauss S Terashima K 2014 *Plasma sources Sci. Technol* **23** 025019
26. Semnani, A. Venkattraman, A. A. Alexeenko, D. Peroulis: *Applied Physics* Letters 102 (2013) 174102
27. M. Radmilovi´c-Radjenovi´c, Z. Lj. Petrovi´c: Eur. Phys. J.D 54 (2009) 445
28. Meyerand R G. Haught A F. 1963 *Phys. Rev. Lett.* **11** 401-403
29. Raizer Y P. 1991 Gas Discharge *Physics Springer*, Berlin
30. Lisovskiy V A. Yakovin S D. Yegorenkov V D. 1994 *J. Phys. D: Appl. Phys* **27** 2340-2348,
31. Smith H B. Charles C. Boswell R.W. 2003 *Phys. of Plasmas* **10** 875-881
32. Xu N. S. Latham R V. 1995 High Voltage Vacuum Insulation: Basic Concepts and Techno-logical Practice Academic Press
33. Langmuir I. 1928 *Proceedings of the National Academy of Sciences of the United States of America* **14** 627

Environment friendly pervious concrete for sustainable construction

S. Dash [1*], B. Kar[1]

[1]Department of Chemistry, Kalinga Institute of Industrial Technology (Deemed University), Bhubaneswar, Odisha, India-751024

ABSTRACT

Now-a-days our modernized cities are covered with air-water proof building material. It obstructs the lack of air permeability and water permeability common concrete pavement so that the rain water is not filtered underground. A large amount of rain water ends up falling on impervious surface such as parking lots, drive ways, sidewalks and streets rather than soaking into soil. This creates a natural imbalance in the ecosystem and leads to various problems like soil erosion, floods, ground water depletion. A simple solution is to be avoiding these problems to stop construction impervious surface and switch to pervious concrete. Working on rain-drain concept, porous concrete allows large amount of water in the body system resulted ground water rechargement and control storm water management. Pervious concrete pavement is the best solution for protecting trees in a impervious surface. Many plants have faced difficulty growing in impervious because air and water can't touch to the roots. Porous concrete helps the adjacent trees to receive more water and air from the soil. Pervious concrete creates opportunity for lands caper and architects who wish to use greenery in parking lots and paved urban areas. Inspired by the pervious concrete technology and the fact that fly ash is a waste material abundantly available, the research Group at KIIT University, India commenced a comprehensive research programme on fabrication of Fly ash based pervious Material [FPC] from fly ash. The compressive strength of the material falls on the range of 3-28 MPa , porosity 15 - 35 % with permeability 8 - 20 mm/s. This material has potential application such as ground water rechargement, storm water management, noise reduction, control surface run-off, temperature behaviour and pollution retention sinks. The research mainly focuses on the manufacture of fly ash-based pervious material for ground water rechargement and large amount of utilization fly ash.

Keywords:-Pervious, Concrete, Fly Ash, Strength, Permeability, Porosity

Corresponding Address: E-Mail.subhakantadash9@gmail.com

IOP Conf. Series: Materials Science and Engineering **410** (2018) 012005 doi:10.1088/1757-899X/410/1/012005

1. INTRODUCTION

Now-a-days our modernized cities are covered with air-water proof and building material. It obstructs the lack of water and air permeability inside the body system, so rain water is not able to filter underground. A large amount of rain water are not to be utilized in proper manner it ends up on falling drive way, parking lots,pedistrain road rather than soaking in the soil which on further creates some natural imbalance leads to flood, soil erosion,water depletion. A prominent solution of the problem is to controlled impervious surface and switch off to pervious concrete which offers multidirectional benefits [1-3]. Working on rain-drain concept, porous concrete allows large amount of water in the body system resulted ground water rechargement and control storm water management[4-6]. Implementation of porous concrete is one of the ideal solutions for save the trees in a impervious surface. It seems that in impervious surface many plants have not able to grow properly as water and air can't able to touch the root of the tree.[7-9].According to National Ready Mixed Concrete Association (NRMCA, 2004), porous concrete (PC) is a mixture of aggregate, cement, water and admixture combined to form a no-fine structural material. Pervious concrete has many different names including zero-fines concrete, pervious concrete and porous concrete. It is a special highly porous concrete, basically used for concrete flatwork applications. It allowed water pass through the body system and percolate to the ground[10-11]. In the past 30 years, the demand of this concrete is gradually increasing in the United States and Environmental Protection Agency(EPA) recommended it is the Best Management Practices (BMPs).The use of pervious concrete in building site design can also aid in the process of qualifying the building for Leadership in Energy and Environmental Design (LEED) Green Building Rating System credits[12-14].This material has high volume of voids inside the body system which causes lower strength. In pervious concrete, the large sized pores ranging from 2 - 8 mm, porosity ranges from 10- 35% permeability rate from 5-20 mm/s, compressive strength range from 3-30 MPa and density 1600-2000 kg/m^3 [15].

2. BACKGROUND OF PERVIOUS CONCRETE

Pervious concrete was first introduced in the 1800s in Europe as pavement surface and bearing walls. Previously, in the 19th century this concrete was used in a variety of purposes such as load bearing walls, panels and paving. The use of this material is very limited because of its lower strength. Pervious concrete has been extensively used as a structural building material in Europe, Australia and the Middle East for over 70 years. It became popular again in the 1920 for two storey homes in Scotland and England. It became more popular in Europe after the Second World War due to the shortage of cement. It did not become as popular in the US until the 1970s.In India it became popular in 2000. In the United Kingdom in 1852, two houses were constructed using gravel and concrete. It wasn't until 1923 when pervious concrete resurfaced as a usable construction material. In that time it was applied in a limited way to the construction of two story homes in areas such as Scotland, Liverpool, London and Manchester [ACPA, 2006]. After the World War II era the use of no-fine concrete in Europe increased steadily. Since pervious concrete contains less amount of cement content than conventional concrete and cement was scarce at the time, so pervious concrete was gain the popularity for best material in that period. Once again housing construction was its primary use. Later, pervious concrete spread to areas such as Venezuela, West Africa, Australia and Russia.

3. MATERIALS AND MIX PROPORTION

The specimens are obtained by the mixture of cement, aggregates, little amount of sand or no sand, water and admixture. During the mixing process, extreme care was taken in order to maintain water content so that the pervious concrete can get better workability. Generally ,the mix design were produced with cement to aggregate ratio (C/A) of 1:3, 1:4,1:6 and water cement (w/c) ratio of 0.30,0.35,0.40 respectively are summarized in Table . Additionally admixture 7-15 % of silicafume and 5-7 % of super plasticizer (weight of cement) were added to the mixture in order to get maximum strength.All the mix designs are prepared to achieve good permeability and porosity.The mix proportion of pervious concrete is shown in table 1.

Table 1.Mix proportion of Pervious concrete

Mix Proportion of Pervious Concrete	
Materials	Proportion
Cementious material	270 to 415 Kg/m^3
Aggregates	1190 to 1480 Kg/m^3

Water: Cement ratio(by mass)	0.27 to 0.40
Aggregate: Cement ratio(by mass)	4 to 6:1
Admixture	5 -7 %
Fine: Coarse aggregate ratio (by mass)	0 to 1:1

4. CHARACTERSTICS OF PERVIOUS CONCRETE

The properties of Pervious concrete (PC) such as compressive strength, split tensile strength, permeability, porosity, density, were found vary satisfactory performance and the test results are discussed below in table.2.

Table 2.Properties of Pervious concrete

Properties of Pervious Concrete	
Density	1600 to 2000 Kg/m3
Compressive Strength	3.5 to 28 MPa
Flexural Strength	1 to 3.8 MPa
Tensile Strength	1 to 3 MPa
Permeability	8-20 mm/s
Porosity	15 to 35%
Slump	Zero

5. APPLICATION OF PERVIOUS CONCRETE

Natalia I. Vázquez-Rivera et al.(2015)have studied Optimization of pervious concrete containing fly ash and iron oxide nano particles and its application for phosphorus removal. The laboratory prepared pervious concrete can remove phosphorus with the first-order removal constant at 0.031 h^{-1} and the Freundlich isotherm constant at 2.48 $mg^{1-1/n}$ kg^{-1} $L^{1/n}$ [16].Darshan S. Shah et al.(2013) have studied the application of Pervious concrete : New Era for Rural Road Pavement. This paper gives the idea about uses of pervious concrete in rural road pavement as well as solves ground water depletion and agriculture related problem.Porous concrete is more favourable because it allows water pass in its body system and control storm water runoff. The research concluded that it is possible to prepare 1m × 1m× 0.15m size pavement with durability properties with a minimum case of Rs 29 per m3 or Rs 193 per m^2 or Rs. 18 per $feet^2$ [17]. Shihui Shen et al.(2012) have studied the Pervious concrete with titanium dioxide as a photo catalyst compound for a greener urban road environment In this study titanium dioxide (Ti02) applied into pervious concrete pavement to remove some of these organic pollutant from the air, so that pervious concrete pavement for sustainable application such as air pollutant removal and storm water management [18].M.Carsana et al. (2013) have studied the use of porous concrete as a building material subjected to its strength durability properties and corrosion control through embedded steel. In the present research mechanical as well as durability properties were studied. Apart from this carbonation induced corrosion properties of the concrete were studied. It is observed that w/c ratio 0.34-0.41 and cement to aggregate ratio 1:4 subjected to fast carbonation and it can't provide long term passivation to embedded steel [19]. Morgenroth et al.(2013) ,have studied the belowground effects of porous pavements—Soil moisture and chemical properties. In the present study, how porous pavement related to soil moisture and their chemistry subjected to relate urban vegetarian were studied in Christchurch, New Zealand. The test result indicates that soil PH 5.6 which is acidic becomes going to neutral 6.3 after passing through pavement altered soil. Similarly concentration of soil Fe, Al, and mg decreased[20]. Ravindra Rajah et al.(2010) have studied the environmentally friendly pervious concrete for sustainable construction. This paper reports an experimental investigation into the, physical and engineering properties of pervious concrete having varying amount of low calcium fly ash as the cement replacement material. Replacement of 50% cement by fly ash had no significant effect on water permeability but it was noted that there is a marginal strength effect of pervious concrete. Three previous concrete mixtures were prepared by replacing 0, 20 and 50% of fly ash, and its properties were studied. Based on the data, it is obtained that there is a co-relation between strength and porosity and between permeability and porosity. It also found that pervious concrete maintain a porosity range of 15% to 30%. Also it is assuming that replacement of 50% of cement has no significant effect on water permeability. So it is possible to prepare environment friendly pervious concrete with significantly reduced amount of Portland cement with fly ash [21].Brian Shackel et al.(2006) have studied the Design of permeable paving subject to traffic. This paper describes the multidirectional benefits of porous concrete on conventional concrete in terms of sustainability and environmental impact [22].Norbert Belatte et al.(2010) have studied the Sustainability benefits of pervious concrete pavement. Portland cement pervious concrete is a material of increasing interest for parking lots and other application because of its important sustainability benefits. Pavement concrete greatly reduces the quantity of runoff and first flush pollution from parking areas and can enhance ground water recharge. Also it's found that light colour pavement material can help mitigate urban heat Island effect [23]. Sonia Rahman et.al (2014) has studied the long term drainage performance of no-fines concrete pavements in Canada. It includes the permeability evaluation, performance analysis, and strength assessment [24].

6. LIFE CYCLE COST OF PERVIOUS CONCRETE

No-fine concrete is the ideal solution for light weight roadway such as pedestrian road, parking lots and residential street. As porous concrete is porous is nature, so its various mechanical properties are weaker than conventional concrete but it can solve many environmental issues such as storm water management and het island effect. The best way to compare this material with other is of its cost parameter and application. In the present there is no long studied has been conducted for its cost analysis. If available it is only suitable for local condition. Only few studied have been done on comparative analysis of cost construction between porous concrete and conventional concrete. The result reveals that the initial cost of porous concrete is high due to controlled design and to maintain proper void.If life cycle cost is determined than its overall benefits can be calculated. It is sometimes very difficult to determine the life cycle cost because of lack of large scale testing, performance analysis and maintenance cost. By determining the actual life cycle cost its cost would be reduced up to 30%.In another study it reveals that by installation of pervious concrete it can save up to $64,649 for water treatment benefits and $3,788,856 for installation cost for more than 25 years as compared to conventional concrete. The reduction cost of pervious concrete can be increased because it doesn't require an drainage system. Since more pores present inside the body system helps to percolate water. According to Federal Highway

Administration (FHWA) the cost of Pervious concrete is 15-20 % more than traditional concrete. It basically depends upon few factors such as material employed and application method. In the present report the cost analysis of full, partial and no filtration pavement is studied. It is found from the literature only few work has been done on cost analysis of pervious concrete (PC).And the date available which shows that the construction cost of PC more expensive than other concrete. Due to non-availability of long term performance data. Most of the studied are conducted on storm water runoff but none one include the scope of taking UHI mitigation benefits. The author identify this is a potential area of research to study its various properties such as durability, construction cost, life cycle etc [25-26].

7. ONGOING RESEARCH AT KIIT UNIVERSITY, ODISHA, INDIA

Under the aegis of National Aluminium Company (NALCO, India) project ,the Research Group at KIIT University, Bhubaneswar, Odisha, India commenced a comprehensive research programme on fabrication of Fly ash based pervious Material [FPC] from fly ash. This material has potential application such as ground water rechargement, storm water management, noise reduction, control surface run-off, temperature behaviour and pollution retention sinks. The research mainly focuses on the, the manufacture of fly ash-based pervious material for ground water rechargement and large amount of utilization fly ash. A wide range of pervious concrete mixture samples were prepared at the KIIT University, Bhubaneswar laboratory facilities, which included cube sample of (150*150*150 mm) size were used to measure compressive strength, beam size (500*100*100 mm) were used to measure flexural strength and cylinder size (150*300 mm) were used to measure water permeability as shown in Figure 1.

Figure1.Concrete Sample used in the research

As the demand of pervious concrete gradually increasing day by day ,so various design of pavement materials are prepared in laboratory scale and which can be implemented in parking lots and other areas which shown in Figure 2.

Figure 2.Fly ash based Pavement material

8. CONCLUSIONS

This paper looked at various studies conducted on permeable pavement systems and their current application. Also discussed about the detailed design of permeable interlocking concrete pavement in brief. Maintenance and water quality control aspects are also discussed.. The water quality aspects as well as the recent innovations of PC were highlighted and explained their potential application in various field and explain the possibility of further research work .The recent innovations like water treatment and recycling pavement system, development of a combined geothermal heating and cooling is promising and it is detailed in cut short, future research works are outlined in brief. These permeable pavement systems are changing the way human development interacts with the natural environment. Its application towards parking lots, highways and even airport runways are all improvements in terms of water quality, water quantity and safety.

9. REFERENCES

[1]H. Li, J. Harvey, T. J. Holland, and M. Kayhanian, 2013, Environmental Research Letters, Vol. 8 (1), pp. 015023-015036.

[2]H. Li, D. Jones, and J. Harvey, 2012, Transportation Research Record: Journal of the Transportation Research Board, Vol. 2305, pp 83-94.

[3]M. Kayhanian, D. Anderson, J.T. Harvey, D. Jones, and B. Muhunthan, 2012, Journal of Environmental Management, Vol. 95(1), pp. 114-123.

[4]S. Kosmatka,B. Kerkhoff, and W. Panarese, 2002 Design and control of concrete mixtures, 14th Edition, Portland Cement Association 358.

[5]S. Dhase, P Kumari, and L. J. Bhagia, 2008, Journal of Scientific and Industrial Research, Vol. 67, pp. 11-18.

[6]Alam J and Akhtar M.N 2011 Fly ash utilization in different sectors in Indian scenario ,Issue 1, ISSN 2249-6149.

[7] D. C. Adriano, A. L. Page, A. A. Elseewi, A. C. Chang[2] and I. Straughan, 1979, Journal of Environmental Quality, Vol. 9 No. 3, pp. 333-344.

[8]C. Rajasekhar, 1995, Retention and permeability characteristics of clays and clay fly ash systems subjected to flow of contaminants, Ph.D. Thesis, Indian Institute of Science, Bangalore.

[9]Watanabe Sato Co., Procedure of eco paste for porous concrete, 2005, pp. 1-9.

[10]National Ready Mixed Concrete Association(NRMCA), 2004 What, Why, and Pervious Concrete, Concrete in Practice series, CIP 38, Silver Spring, Maryland 2.

[11]ACI (American Concrete Institute), 2010, Report on Previous concrete, ACI, Farmington Hills, MI USA, ACI 522R-10.

[12]A. Torres, J. Hu, and A. Ramos, 2015, Construction and Building Material, Vol. 95, pp. 850-859.

[13]C. Lian and Y. Zhuge, 2010, Construction and Building Material, Vol. 24, pp. 2664-2671.

[14]M. Gesoglu, E. Goneyisi, G. Khoshnaw, and S. Ipek, 2014, Construction and Building Materials. Vol. 73, pp. 19-24.

[15]P. Chindaprasirt, S. Hatanaka, N. Mishima, Y. Yuasa, and T. Chareerat, 2009, International Journal of Mineral Metallurgy and Materials. Vol. 16, pp. 714-719.

[16]I. Natalia, 2015, Construction and Building Materials, Vol. 93, pp. 22-28.

[17] D.S. Shah, and J. Pitroda, 2013, International Journal of Engineering Trends and Technology, Vol. 4, pp. 3495-3498.

[18]S. Shen and M. Burton, 2012, Construction and Building Materials, Vol. 35, pp. 874-883.

[19]M. Carsana, F. Tittarelli, and L. Bertolini, 2013, Cement and. Concrete Research, Vol. 12, pp. 64-73.

[20]J. M.G. Buchan, and B.C. Scharenbroch, 2013, Soil moisture and chemical properties, Vol. 51, pp. 221-228.

[21]R. Sri Ravindrarajah, and A. Yukari, 2010, environmentally friendly pervious concrete for sustainable construction, 35th Conference on OUR WORLD IN CONCRETE & STRUCTURES: 25 – 27 August, Singapore.

[22]B. Shackel, 2006, Design of permeable paving subject to traffic, 8th International Conference on Concrete Block Paving, November 6-8, 2006 San Francisco, California USA.

[23]N. Delatte and S.S. Schwartz, Sustainability, 2010, Benefits of Pervious pavement, Second International conference on sustainable construction materials and technologies June 28-30, Universita politechnic, delle Marche, Ancona, Italy.
[24]S. Rahman, V. Henderson, L. Susan, S. L. Tighe, 2014, Long Term Drainage Performance of Pervious Concrete Pavements in Canada, Paper prepared for presentation Design, Construction and Maintenance of Permeable Pavement Session Conference of the Transportation Association of Canada Montreal, Québec, pp. 1-16.
[25]T. Wang, J.T. Harvey, D. Jones, 2010, A Framework for Life-Cycle Cost Analyses and Environmental Life-Cycle Assessments for Fully Permeable Pavements, Technical memorandum, Institute of Transportation Studies, Caltrans document No.: CTSW-TM-09-249.03 UCPRC document No.: UCPRC-TM-2010-05.
[26] S.L. Terhell, K. Cai, D. Chiu, and J. Murphy, 2015, Cost and Benefit Analysis of Permeable Pavements in Water Sustainability, pp1-8.

Temperature Sensitive Conductivity Studies of NBT-Based Lead-Free Ternary System

S. Praharaj and D. Rout*

School of Applied Sciences, KIIT Deemed to be University, Bhubaneswar, Odisha, 751 024, India
*Corresponding author: E-mail: droutfpy@kiit.ac.in, dibyaranjanr@gmail.com

Abstract.

The frequency and temperature dependence electrical conduction characteristics of a lead-free 0.78NBT-0.2ST-0.02BT ternary system were investigated by impedance spectroscopy. Variation of real part of conductivity with frequency reveals a frequency independent and a dispersive region obeying Jonscher power law. The relatively higher values of S-parameter in the mid temperature range envisage the presence of localized relaxation or hopping of the charge carriers. The activation energies calculated from the temperature dependent hopping frequency and dc conductivity, portray the increase of unsuccessful hops of the charge carriers with increase in temperature according to the Jump relaxation model. Interestingly the conductivity values could be curbed by introduction of BT in to NBT-ST. The collapsing of the scaled conductivity spectrum in to a single master curve depicts the temperature independent conduction relaxation mechanism.

Keywords: Conductivity, Ternary system, activation energy, relaxation mechanism

1. INTRODUCTION

Relaxor ferroelectrics (RFEs), particularly $Na_{0.5}Bi_{0.5}TiO_3$ (commonly abbreviated as NBT) has been the center of interest in the electronic industries to cater the urgent surge for creating a lead free environment. Extensive research has brought in the use of these ceramics as sensors, actuators, ultrasonic transducers, multilayer capacitors, non-volatile memories etc., owing to their adequate electromechanical properties along with peculiar phase transition and relaxor behaviour [1]. However, NBT comes with some detrimental properties like large coercive field and high conductivity which makes the poling process difficult. During high temperature sintering, possibly the volatilization of Bi^{2+} ions along with partial reduction of Ti^{4+} ions to Ti^{3+} incorporates oxygen vacancies, which may be responsible for its large conductivity [2, 3]. In this context, solid solutions of NBT with other perovskites like $BaTiO_3$ (BT), $SrTiO_3$ (ST), $K_{0.5}Ba_{0.5}TiO_3$ (KBT), KNO_3 (KN), $K_{0.5}Na_{0.5}NO_3$ (KNN) [4-8] are considered among which the rarely investigated species NBT-ST emerges as a potential candidate for lead free alternative to widely used PZT in piezoelectric applications. Substitution of Sr^{2+}

ions found to be suitable to improve the relaxor properties in NBT. Further, introduction of BT into NBT-ST enhances its piezoelectric properties forming a noteworthy ternary system. Recently, studies on piezoelectric and dielectric properties of NBT-ST-BT has been carried out by Praharaj et al. [9, 10]. Nevertheless, a special attention on conductivity study of the samples is highly desirable to facilitate the ease of poling process. From such a viewpoint, an attempt has been made to probe into the conductivity of 0.78NBT-0.2ST-0.02BT (NBT-ST-2BT) ternary solid solution and improve its electrical properties. Impedance spectroscopy has been used as an effective tool to explore the conductivity relaxation of the sample.

2. EXPERIMENTAL DETAILS

2.1. Materials and Method

Dense ceramic samples of $0.78Na_{0.5}Bi_{0.5}TiO_3$-$0.2SrTiO_3$-$0.02BaTiO_3$ with 100% perovskite phase were prepared from Na_2CO_3 (99.8%, Acros Organics, New Jersey, USA), Bi_2O_3 (99.9%, Kojundo Chemical Lab Co., Saitama, Japan), $SrCO_3$ (99.9%, Aldrich), $BaCO_3$ (99.98%, Sigma-Aldrich, St. Louis, USA) and TiO_2 (99.9%, Sigma-Aldrich, St. Louis, USA) by high temperature solid state reaction method. The details of sample preparation has been given elsewhere [9].

2.2. Characterization Techniques

The impedance measurements were carried out in a PSM1735 Impedance Analyser (N4L-NumetriQ) at a frequency range of 10^{-3} – 10^{3} kHz and by using a programmable furnace in the temperature range of 50 – 650 °C. Prior to this, the sintered pellets were sectioned and polished to a 0.25 μm finish. Both the surfaces of the samples were electroded with silver paste and then annealed at 500 °C for 2 hrs.

3. RESULTS AND DISCUSSION

Fig. 1(a) and its inset representing the Z' (*f*) spectra of NBT-ST-2BT in the temperature range 50-650 °C follow a sigmoidal pattern. As indicated in the figure, the Z' values decrease with the increase in temperature in the low frequency regime and forms a plateau like region from 400 °C onwards. A decreasing trend in the magnitude of Z' with temperature clearly indicates the increase of conductivity which may be possibly attributed to the formation of oxygen vacancies with rising temperature to maintain the charge neutrality. This is followed by high frequency tail or saturation region which may be due to the lowering of barrier height at high frequencies and release of space charge increasing the ac conductivity of the system. Z"(*f*) spectra shown in Fig. 1(b) and its inset in the same temperature range also shows a marked dependence on temperature. The plot at all temperatures consists of a low frequency peak (≥ 400 °C) and a high frequency tail. It also displays a decrease in the Z''_{max} values along with an asymmetric broadening of the peaks and a shift in the inflation frequency (frequency corresponding to Z''_{max}) to higher frequency side with the rise in temperature. Such facts suggest temperature sensitive relaxation processes and active conduction mechanism [3]. Hence to probe into the electrical properties of the system under investigation, a meticulous survey of the conductivity performance is the sole aim of this piece of work. Fig. 2 illustrates the frequency dependence of real part of conductivity from 50 °C to 650 °C. Interestingly, the values of conductivity obtained are much less than that reported earlier for NBT-ST samples [11] which may be attributed to the introduction of Ba^{2+} ions at the A-site of the perovskite compound.

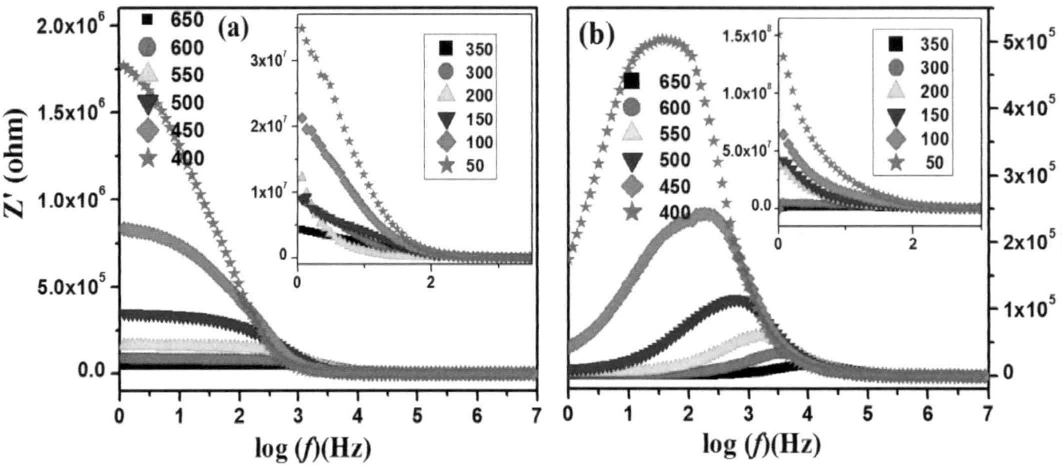

Figure 1. Frequency dependent Z' and Z'' of NBT-ST-2BT at different temperatures

At all temperatures, the plot consists of a nearly frequency independent region corresponding to dc conductivity (σ_{dc}) segregated from a frequency dispersive region. It can be noted that the frequency at which dispersion gains prominence shifts towards the higher frequency side as the temperature increases. The onset of conductivity relaxation is indicated by a crossover from frequency independent region to frequency dispersive region and is analyzed by using Jonscher power law [12]:

$$\sigma(\omega) = \sigma_{dc} + A\omega^S \quad (1)$$

where σ_{dc} is the dc conductivity, S-parameter represents the degree of interactions between the mobile charge carriers with the surrounding lattice lying between 0 and 1; both A and S are weakly temperature dependent. Values of the S-parameter at all temperatures lie around 0.9 which depicts the interactive nature of the charge carriers. A closer introspection of the S-parameter (Fig. 3(c)) reveals that in the mid temperature range the values are relatively more than the high and low temperature zone.

This suggests localized relaxation associated with the charge carriers (polar nano regions as NBT-ST-2BT is a RFE [9]) rather than long range barrier hopping in the mid temperature range [3]. The non-existence of polar nano regions in the high temperature regime which are almost frozen at low temperatures give rise to long ranged conduction relaxation. The frequency response of the electrical conductivity may be explained by the jump relaxation model [13]. According to this model, at lower frequencies of the ac field, there is a high probability of the charge carriers hopping to their neighboring sites and staying there since the neighborhood has enough time to be relaxed with respect to the new position of the charge carriers. The conductivity in the low frequency region is related to the successful hopping mechanism and gives rise to the dc conductivity. However, beyond the low frequency region, the number of unsuccessful hops increases due to higher frequencies of the ac field. Such a transformation arises at a frequency called the hopping frequency (f_H). In the present sample, f_H is calculated using the formula [14]:

$$f_H = (\sigma_{dc} / A)^{1/S} \quad (2)$$

The f_H is found to be temperature dependent and obeys the Arrhenius law (Fig. 3(a)):

$$f_H = f_0 e^{-(Ea/kT)} \quad (3)$$

where E_a is the activation energy for the successful hopping.

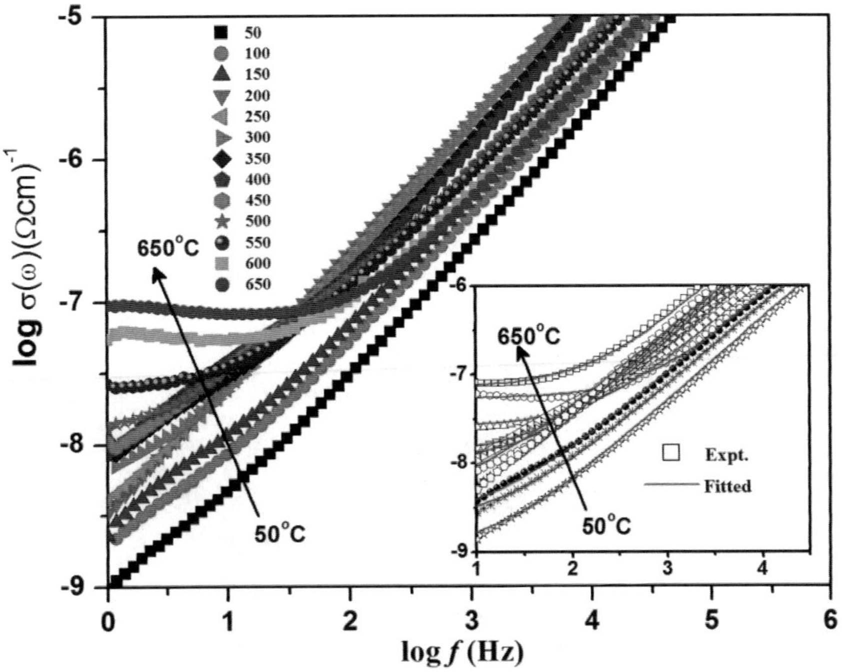

Figure 2. Frequency dependent ac conductivity of NBT-ST-2BT at different temperatures

The plot separates into two distinct regions of different activation energies bridged by a flat region. The flat region may associate with the temperature range over which the nucleation of PNRs takes place. Values of E_a obtained enunciate the fact that the number of success hops is more at low temperatures. At higher temperatures, the flipping frequencies of the charge carriers increase leading to more unsuccessful hops. To further support this fact, log σ_{dc} vs 1000/T [Fig 3(b)] obeying Arrhenius law:

$$\sigma_{dc} = \sigma_0 \, e^{-(E_{dc}/kT)} \tag{4}$$

where E_{dc} is the activation energy for dc conduction. The higher value of E_{dc} in the high temperature interval gives an impression that the frequency dispersion increases as the temperature increases. Further it is also found that values of E_a and E_{dc} are close to each other. To have a better insight into the common underlying physics of the conductivity mechanism, scaling studies have been conducted and at different temperatures, the real part of conductivity following [15]:

$$\sigma(\omega) / \sigma_{dc} = F(f / f_0) \tag{5}$$

where f_0 is the scaling frequency, F is the scaling function independent of temperature. Fig. 3(d) demonstrates the scaled conductivity spectra which collapsed into one master curve at temperatures which clearly depicts nearly temperature insensitive conduction mechanism though a little dispersion is found on the higher frequency side.

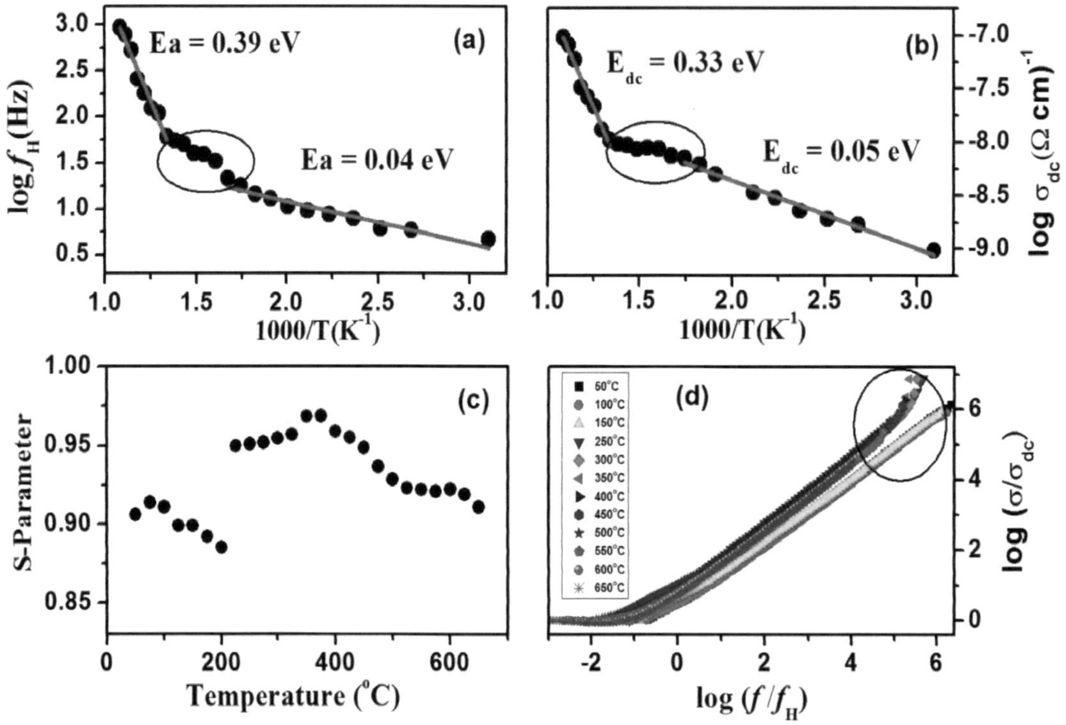

Figure 3. (a) log f_H vs. 1000/T, (b) log σ_{dc} vs. 1000/T, (c) S-parameter vs. Temperature and (d) conductivity scaling curves of NBT-ST-2BT at different temperatures.

4. Conclusion

NBT-ST-2BT ceramics prepared by conventional solid state reaction technique show an increase in the conductivity with temperature as evidenced from the Z'(*f*) and Z"(*f*) spectra. The variation of real part conductivity with frequency which consisted of a nearly frequency independent region and a frequency dispersive region followed the universal power law. The higher values of S-parameter in the mid temperature range symbolize conductivity due to localized hopping or short ranged ordered orderings. The temperature dependence of hopping frequency and dc conductivity obey Arrhenius law and activation energies calculated from both the fittings segregated into two distinct regions with a flat and temperature insensitive region in between them. Finally, the scaling studies on conductivity reveal almost temperature insensitive conduction mechanism due to the collapsing of the all the curves into a single master curve. It is observed that the conductivity can be pinned down by the introduction of BT in to NBT-ST system furnishing an easier poling process of the sample.

Reference

[1] J. Rodel, W. Jo, K.T.P. Seifert, E. Anon, T. Granzow, and D. Damjanovic, 2009, *Journal of American Ceramic Society, Vol.* 92, pp.1153.
[2] T.R. Shrout and S.J. Zhang, 2007, *Journal of Electroceramics, Vol.* 17, pp. 111.
[3] R. Kumari, N. Ahlawat, A. Agarwal, S. Sanghi, M. Sindhu, and N. Ahlawat, 2016, *Journal of Alloys and Compounds,* Vol. 676, pp. 452.
[4] S.S. Sundaria, B. Kumar, and R. Dhanasekaran, 2013, *Ceramic International,* Vol. 391, pp. 555.

[5] D. Rout, K.S. Moon, S.J.L. Kang and I.W. Kim, 2010, *Journal of Applied Physics, Vol.*108, pp. 084102.
[6] Babu G A, Raja R S, Bhaumik I, Ganesamoorthy S, Ramasamy P and Gupta P K 2015 *Material Research Bulletin,* Vol. 53, pp. 136.
[7] Jiang X, Wang B, Luo L, Li W, Zhou J and Chen H 2014 *Solid State Chemistry, Vol.*213, pp. 72.
[8] P. Laoratanakul, R. Yimniun, and S. Wongsaenmai, 2011, *Current Applied Physics,* Vol. 11, pp.161.
[9] S. Praharaj, D. Rout, V. Subramanian and S.-J. L. Kang, 2016, *Ceramic International,* Vol. 42, pp. 12663.
[10] S. Praharaj, D. Rout, S.-J. L. Kang and I.W. Kim, 2016, *Materials Letter, Vol.* 184, pp. 197.
[11] A.R. Malathi, G.S. Kumar and G. Prasad, 2015, *Ferroelectrics,* Vol. 481, Vol. 21.
[12] A.K. Jonscher, 1977, *Nature, London,* Vol. 264, pp. 673.
[13] K. Funke, 1992, *Progress Solid State Chemistry,* Vol. 29, pp.11.
[14] B. Roling, A. Happe, K. Funke, and M.D. Ingram, 1997, *Physics Review Letter,* Vol. 78, pp. 2160.
[15] P. Singh, O. Prakash, and D. Kumar, 2011, *Physics Review B, Vol.* 84, pp. 174306.

Effect of different fibers on compressive strength of self-compacting concrete at elevated temperature

Mohammad Zuhair[a], S. K. Deshmukh [b],

[a]P. R. PotePatil College of Engineering & Management Amravati,
[b]College of Engineering & Technology, Akola, Maharashtra, India,
Corresponding Email.: zuhairgcoe@gmail.com and principalcoeta@gmail.com

ABSTRACT

Self-compacting concrete (SCC) is a highly flow able type of concrete that spreads into the form without the need of mechanical vibration and placed by means of its own weight. It is necessary to know the change in the properties of concrete due to extreme temperature exposure. This experimental study investigates the effect of various fibers used in Self Compacting Concrete (SSC) on compressive strength when exposed to elevated temperature from 400^{0}C, 600^{0}C and 800^{0}C for the duration of 2 hrs. and 4 hrs. The Self Compacting Concrete (M30 grade) with 1% and 0.5% of different fibers (Steel, Glass, and Polypropylene) were used. The results indicate the increase in temperature and heating duration decreases compressive strength of concrete. Elevated temperature upto 800^{0}C declines the performance of concrete comparing the same with lower temperature of 400^{0}C. The use of 1% Steel fiber and 0.5% Polypropylene fiber shows significant performance in the development of higher compressive strength and improved toughness at elevated temperatures and addition of fibers reduces the spalling in concrete.

Keywords—*Compressive strength, Elevated temperature, Steel fiber, SSC,Polypropylene fiber.*

1. INTRODUCTION:

1.1 General:

Self-compacting concrete, a concrete with high flow ability that can be placed and compacted under its own weight without any external vibration, allows complete filling of formworks along with the complete covering of the reinforcing bars in compact matrix structure. SCC has many advantages that include faster construction, better surface finishes, easier placing, reduction in noise levels and improved durability but it affects when exposed to fire and loses its strength.

The rapidly advancing technology of structural design and fire safety created a need for more precise information on the behavior of structures during fires. The maximum temperature reached during a fire is normally estimated indirectly from the melting of ingredients of concrete. A temperature of 1000ºC to 1100ºC observed during fire in residential buildings for duration mostly found to be in between 1 to 2 hours. For public buildings, temperature rises up to 1100ºC to 1200ºC and the fire duration may exceed 2 to 3 hours in some cases. Still higher temperatures have been observed during fires in industrial buildings and ware houses in which considerable quantities of solid and liquid lubricants are stored. The fire in industrial buildings may last for more than two hours and the temperature may go beyond 1300ºC. Among these temperature of 1000 to 1100ºC during fires which lasted for a duration of 1 to 2 hours has been observed more frequently. Thus, temperature and its duration affects exposed concrete, it chips and flakes known as "spalling"-resulting from the vaporization of water trapped within due to the high temperature. In concrete structure, chips split away from Ceilings, walls and supporting pillars, reducing their load bearing capacity and increasing the risk of collapse. Performed experimental study on Carbonate, siliceous, and lightweight aggregate concrete and were heated for short duration to temperatures of 200 to 1600° F (93 to 871°C).The lightweight concrete had strength characteristics at high temperatures similar to those of carbonate concrete. At temperatures above 800°F (427°C), the siliceous aggregate concrete had lower strength than the other concretes [1]. HSC lost its mechanical strength in a manner similar to that of NSC. The range between 400 and 800°C was found to be critical to the strength loss [2].

The relationship between occurrence of explosive spalling and residual mechanical properties of fiber such as Steel fiber, polypropylene fiber, and hybrid fiber, toughened high performance concrete exposed to high temperatures. After exposure to high temperatures ranged from 200 to 800 °C, the residual strength was decreased by exposure to high temperatures over 400 °C, residual fracture energy was significantly higher than that before heating. Incorporating hybrid fiber seems to be a promising way to enhance the resistance of concrete to explosive spalling [3]. This study of using various fibers concluded that the use of fibre cocktail reinforced SCHPC material can be very effective in reducing thermal stress and in improving composite effect of hybrid fibres on the post crack behaviour during heating process at high temperature [4].

IOP Conf. Series: Materials Science and Engineering **410** (2018) 012007 doi:10.1088/1757-899X/410/1/012007

1.2 Scope of Study:

The efficiency of SCC can be increase by introducing of fibers like polypropylene fibers, steel fibers and glass fibers, further enhance their toughness, tensile strength, resistance to crack propagation thereby improving the durability properties. The effects of fibers were qualified based on the fiber volume, length, and its aspect ratio, other properties of fibers such as shape and surface roughness are also found to be important. This work contributes for experimental evaluation of compressive strength of SCC with fibers exposed to elevated temperature up to 800^0C for 2 hrs. and 4 hrs. of heating duration. Steel fibers, Glass fibers and Polypropylene fibers are used for this demonstration in two different percentage as 0.5% and 1% as steel fibers consists high elastic modulus and stiffness, glass fibers improve the interface zone properties and consequently the fiber-matrix bond leading to enhanced post-cracking toughness and energy absorption capacity and Polypropylene fibers allow the water vapour to escape without increasing the internal pressure, so the concrete structure remains intact.

2. Experimental Work

Concrete exposed to temperature higher than 300ºC may experience loss in strength. High strength concrete is more susceptible to explosive spalling when exposed to temperatures above 300ºC. This research was designed to study the behavior of self-compacting concrete subjected to elevated temperatures by experimental investigation on compressive strength with glass fibers, Polypropylene fibers and steel fibers.

2.1 Program Work

Figure 2.1: Glass Fibers

To modify the concrete to such a scale that it would provide higher strength even at tremendous temperature and concrete should be self-compacting for the ease of placement at congested reinforcement. To reinforce our idea we have used three different fibers (steel, glass and polypropylene) at three different temperature (400,600 and 800 degree Celsius) for two different time intervals (2 and 4 hours) in muffled furnace.

2.2 Material Properties:

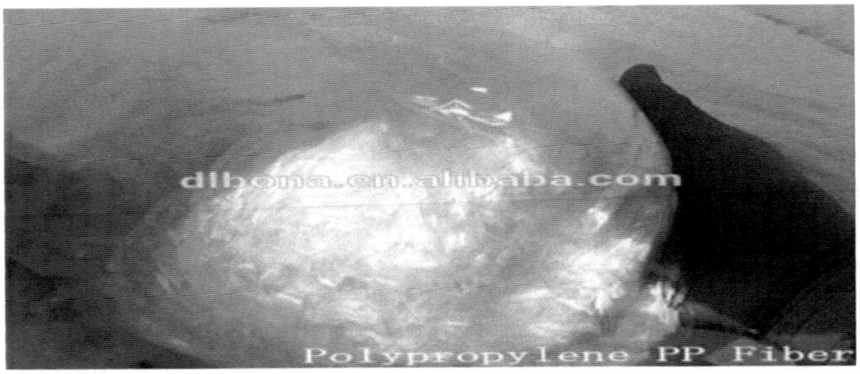

Figure 2.2: Polypropylene fibers

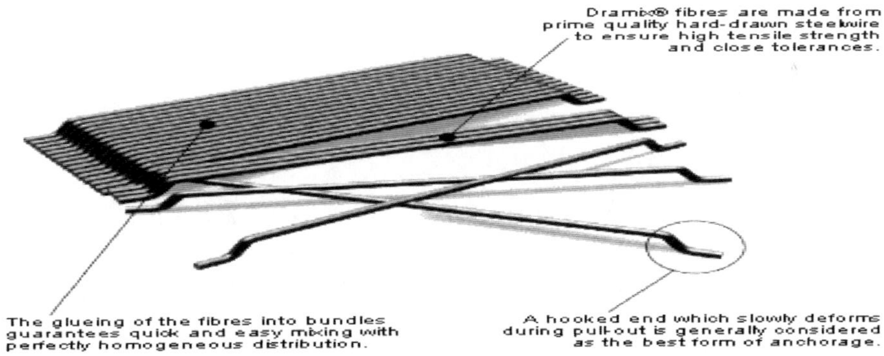

Figure 2.3: Steel Fibers

Characteristic compressive strength (fck) = 30Mpa
Packing factor = 1.04
Specific gravity of cement (Gc) = 3.12
Specific gravity of fine aggregate (Gfa) = 2.61
Specific gravity of coarse aggregate (Gca) =2.68
Specific gravity of water (Gw) = 1.0
Bulk density of FA (Wfa) =1594 Kg/m3
Bulk density of CA (Wca) =1264 Kg/m3
The volume ratio of fine aggregate = 0.54
Super plasticizer (sp.dosage) =0.8 %
Air content = 1.5%

Monofilament type Polypropylene fiber of length 30mm to 50mm.
AR-glass of length 25mm

2.3 Mix Proportion:

Following mix proportion were adopted during the execution, obtained by trial and error method.

Material	Cement	Sand (fa)	Aggregate (ca)	Water
Kg/m^3	430	700.9	524.60	180.6
Ratio	1	1.63	1.22	0.42

2.4 Execution of work:

The fibrous concrete has been classified into three categories depending upon the three types of fiber viz. steel, glass, and polypropylene. All the three fiber have been used in two proportions viz. 1% and 0.5%.. 2 cubes of each proportion is casted for three different temperatures as 400^{0}C, 600^{0}C and 800^{0}C for two different time as 2hrs and 4 hrs. Water cooling adopted after heating in muffle furnace. For each temperature there were two cubes, average of both were taken to avoid any defect in analysis. For the better performance we have used superplasticizer in concrete to enhance its flow ability.

2.5 Performance Analysis:

Figure 2.4: Muffle furnace

It was reported that compressive strength of SCC decreased with increase in temperature. It was found that grade of concrete had an effect on the strength loss of concrete, especially in the temperature range below 400°C. Addition of fibers into concrete improves the overall ductility of the concrete imparting toughness, greater tensile strength, and resistance to fatigue, impact, blast loading and abrasion. The performance of each modified concrete were analyzed based on compressive strength of concrete.

Figure 2.5: Removal of Cube from Muffle Furnace

3. Results and Discussion:

Following are the results represented in terms of compressive strength and its percentage reduction as temperature increases from 400^0C to 800^0C

3.1 Compressive Strength:

a) 1% Fibers for elevated temperature upto 2 hrs:

Table No. 3.1 Compressive Strength of 1% fibers for 2 hrs of elevated tempretaure

Sr. No.	Tempe rature	Steel Fiber	Glass fiber	Polypropelene Fiber
1	400	32.01	27.81	28.13
2	600	32.91	29.36	24.34
3	800	22.92	13.93	13.15

IOP Conf. Series: Materials Science and Engineering **410** (2018) 012007 doi:10.1088/1757-899X/410/1/012007

Table No. 3.1 and Figure No. 3.1 shows tabular and graphical representation of average compressive strength of M30 grade concrete with 1% steel fiber, glass fiber and polypropylene fiber respectively at 400°C, 600°C and 800°Cfor 2 hours, after 7 days curing.

Graph No. 3.1 Compressive Strength of 1% fibers for 2 hrs of elevated temperature

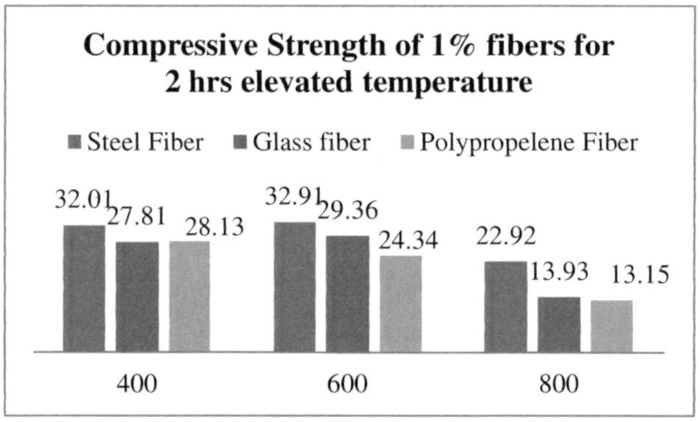

Above observations were noted after 7 days of curing for 1% fibers and exposed to elevated temperature for 2 hrs. It shows, for steel fiber, compressive strength of concrete cubes exposed for 400^0C temperature as 32.01 N/mm^2 but if temperature increased upto 800^0C for the same the value of strength reduced to 22.92 N/mm^2. Similarly, for glass fiber strength decreased from 27.81N/mm^2 to 13.93 N/mm^2 and for Polypropylene fiber its value dropped from 28.13 N/mm^2 to 13.15 N/mm^2.

b) 1% Fibers for elevated temperature upto 4 hrs

Table No. 3.2 and Figure No. 3.2 shows tabular and graphical representation of average compressive strength of M30 grade concrete with 1% steel fiber, glass fiber and polypropylene fiber respectively at 400°C, 600°C and 800°Cfor 4 hours, after 7 days curing.

Table No. 3.2 Compressive Strength of 1% fibers for 4 hrs of elevated temperature

Sr. No.	Temperature	Steel Fiber	Glass fiber	Polypropelene Fiber
1	400	35.4	23.91	19.79
2	600	27.91	24.81	19.79
3	800	12.36	17.75	6.33

Graph No. 3.2 Compressive Strength of 1% fibers for 4 hrs of elevated temperature

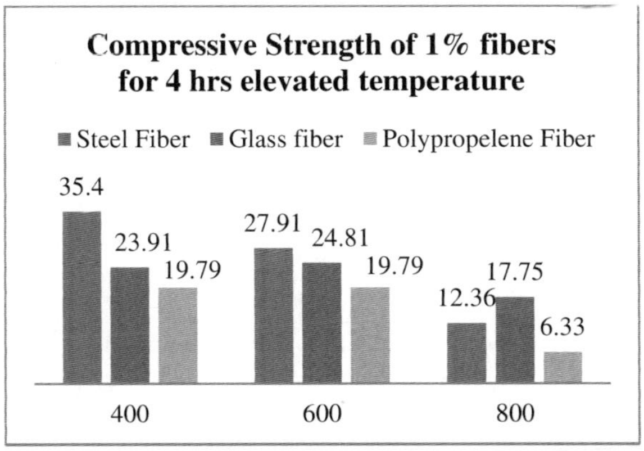

Above results shows the readings noted for SCC with 1% fibers and exposed to elevated temperature for 4 hrs after 7 days of curing. During this experiment it was observed that, concrete with steel fiber exposed for 400^0C temperature achieved compressive strength as 35.4 N/mm^2 but further increase in temperature to 800^0C the strength reduced to 12.36 N/mm^2. Similarly, for glass fiber strength decreased from 23.91N/mm^2 to 17.75 N/mm^2 and for Polypropylene fiber its value dropped from 19.79 N/mm^2 to 6.33 N/mm^2.

c) Comparison of Compressive Strength with 1% Fibers (Steel, Glass and Polypropylene)for elevated temperature 2 hrs and 4 hrs:

Table No. 3.3 Comparison of Compressive Strength with 1% Fibers elevated temperature 2 hrs and 4 hrs

Temperature	Steel Fiber		Glass fiber		Polypropylene Fiber	
	2 Hrs	4 Hrs	2 Hrs	4 Hrs	2 Hrs	4 Hrs
400	32.01	35.4	27.81	23.91	28.13	19.79
600	32.91	27.91	29.36	24.81	24.34	19.79
800	22.92	12.36	13.93	17.75	13.15	6.33

Table above represents the comparative study of compressive strength obtained after using 1% fibers in SCC of grade M30 exposed for two different time duration as 2 hrs and 4 hrs. It shows reduction in compressive strength as duration of exposure temperature increases.

IOP Conf. Series: Materials Science and Engineering **410** (2018) 012007 doi:10.1088/1757-899X/410/1/012007

d) 0.5% Fibers for elevated temperature upto 2 hrs

Table No. 3.4 and Figure No. 3.3 shows tabular and graphical representation of average compressive strength of M30 grade concrete with 0.5% steel fiber, glass fiber and polypropylene fiber respectively at 400°C, 600°C and 800°Cfor 2 hours, after 7 days curing.

Table No. 3.4 Compressive Strength of 0.5% fibers for 2 hrs of elevated temperature

Sr. No	Temperature	Steel Fiber	Glass fiber	Polypropylene Fiber
1	400	29.2	30.69	32.705
2	600	32.44	26.22	25.79
3	800	17.33	17.26	12.46

Graph No. 3.3 Compressive Strength of 0.5% fibers for 2 hrs of elevated temperature

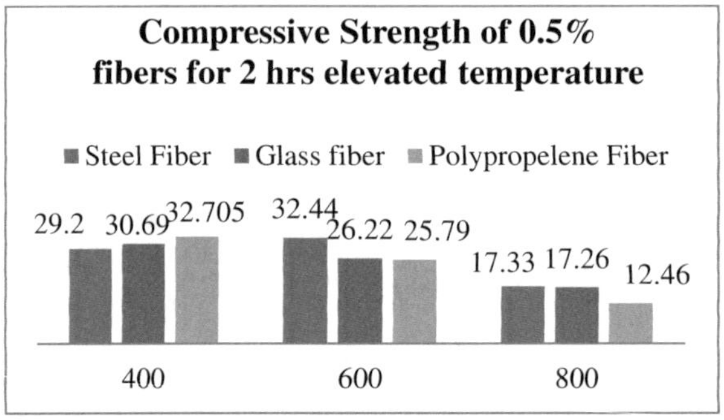

Table and graph above shows the observation obtained for 0.5% fibers and exposure for 2 hrs of elevated temperature for M30 grade SCC. In this case, strength for 400^0C temperature were noted as 29.2 N/mm^2, if further increase in temperature upto 800^0C strength dropped to 17.33 N/mm^2 for steel fibers. For glass fibers same variation has been found, decrease in the value from 30.69 N/mm^2 to 17.26 N/mm^2. Use of 0.5% Polypropylene fiber in M30 SCC, the noted value of compressive strength was 32.705 N/mm^2 under 400^0C and decreased to 12.46 N/mm^2 for temperature 800^0C.

e) 0.5% Fibers for elevated temperature upto 4 hrs

Table No. 3.5 and Figure No. 3.4 shows tabular and graphical representation of average compressive strength of M30 grade concrete with 0.5% steel fiber, glass fiber and polypropylene fiber respectively at 400°C, 600°C and 800°Cfor 2 hours, after 7 days curing.

Table No. 3.5 Compressive Strength of 0.5% fibers for 4 hrs of elevated temperature

SrNo	Temperature	Steel Fiber	Glass fiber	Polypropylene Fiber
1	400	27.97	36.92	29.26
2	600	26.39	21.27	20.74
3	800	13.35	12.52	7.79

Graph No. 3.4 Compressive Strength of 0.5% fibers for 4 hrs of elevated temperature

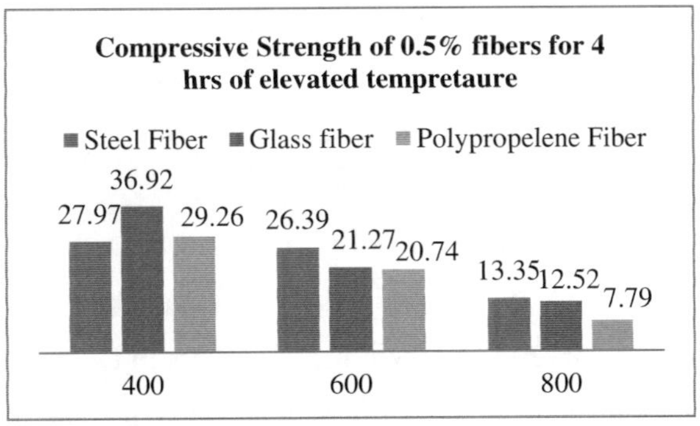

Above table and figure are the tabular and graphical representations of result obtained after using 0.5% of fibers in SCC of M30 grade under elevated temperature of 400^0, 600^0 and 800^0 for 4 hrs. It shows falling in strength of concrete as the temperature increases, it is observed for steel fibers strength falls from 27.97 N/mm^2 to 13.35 N/mm^2, for glass fibers from 36.92 N/mm^{2}to 12.52 N/mm^2 and for polypropylene fibers it decreases from 29.26 N/mm^2 to 7.79 N/mm^2.

IOP Conf. Series: Materials Science and Engineering **410** (2018) 012007 doi:10.1088/1757-899X/410/1/012007

f) Comparison of Compressive Strength with 0.5% Fibers (Steel, Glass and Polypropylene)for elevated temperature 2 hrs and 4 hrs:

Table No. 3.6 Comparison of Compressive Strength with 0.5% Fibers elevated temperature 2 hrs and 4 hrs

Sr No	Temperature	Steel Fiber		Glass fiber		Polypropylene Fiber	
		2 Hrs	4 Hrs	2 Hrs	4 Hrs	2 Hrs	4 Hrs
1	400	29.2	27.97	30.69	36.92	32.7	29.26
2	600	32.44	26.39	26.22	21.27	25.79	20.74
3	800	17.33	13.35	17.26	12.52	12.46	7.79

Table above shows the change in compressive strength as duration of exposure of elevated temperature increases. This change in values shows falling in strength as duration increases for SCC of grade M30 with 0.5% fibers

4. Conclusion

Following conclusions can be drawn from the study work performed experimentally on cubical specimen of M30 grade SCC exposed to elevated temperature from 400^0C to 800^0C in presence of steel fibers, glass fibers and polypropylene fibers in two different ratios.

1. For 2 hrs of heating to specimen with 1% fibers shows the reduction in compressive strength in case of glass fiber and polypropylene fiber and it is also observed that strength of specimen with steel fibers heating upto 800^0C for duration of 2 hrs reduced by 28.46% compare to specimen in 400^0C temperature. The same mix with 1% glass fibers shows 50% falling in compressive strength at 800^0C. The concrete mix with steel and glass fibers gives more strength i.e. 53.25% greater as compared with polypropylene fibers.
2. Increasing in duration from 2 hrs to 4 hrs also suffers compressive strength of same grade of concrete with same percentage of fibers. Exposure to elevated temperature upto 800^0C for 4 hrs decrease the performance of concrete by 64.80%, 25.76% and 68.01% for steel, glass and polypropylene fibers compare to the strength of specimen under 400^0C temperature for same duration of 4 hrs.

3. Comparing the effect of type of fibers in concrete, 1% steel fiber gives the better performance compare to other fibers used in same percent.
4. By reducing percentage of fibers from 1% to 0.5% for 2 hrs of heating for 400^0C, 600^0C and 800^0C shows compressive strength dropped by 40.65%, 43.76% and 62.38% as temperature increases from 400^0C to 800^0C in presence of steel fibers, glass fibers and polypropylene fibers respectively.
5. 0.5 % Polypropylene fibers in concrete exposed to 400^0C temperature for 2 hrs shows better performance compare to glass and steel fibers
6. The specimen with 0.5% of fiber when heated for 4 hours shows varying result in all three fibers with highest resistance to compressive force in steel fiber and least in polypropylene fiber while glass fiber remains at average with least fluctuation.
7. Addition of fibers to self-compacting concrete improves compressive strength of the mix.
8. The tested SCFRC mixtures with its different constitutions did not show a pronounced negative effect on the maturity of concrete, and leads to a higher strength gain as well.
9. After observing all parameters we can say that steel fiber works better at all three temperatures but its fluctuation in compressive strength is quite higher than glass fiber

5. References:

[1] Qiang Du, Jing Wei. "Effects of High Temperature on Mechanical Properties of Polyvinyl Alcohol Engineered Cementitious Composites" Springer, International journal of civil engineering 2017.

[2] M. S. Al-Lami "Effect Of Elevated Temperature On Compressive Strength Of Self Compacting Concrete Using Viscocrete And Silica Fume. International Journal of Civil Engineering and Technology (2017) pp. 405–413

[3] Guanglin Yuan, Qingatao Li. "The Use of surface coating in enhancing the mechanical properties and durability of concrete exposed to elevated temperature" *Construction and Building Materials* 95, (2015): 375-383.

[4] Kulkarni, Yaragal and Narayan , "An Overview Of High Performance Concrete At Elevated Temperatures", *International Journal of Applied Engineering and Technology*. (2011): 48-60.

[5] Anand.N, Prince Arulraj.G, " The effect of elevated temperature on concrete materials - A literature review" *International Journal Of Civil And Structural Engineering* 1, no 4, (2011) 928-938

[6] Srinivasa Rao K, Potha Raju M. and Raju P.S.N "Effect of Elevated temperature on compressive strength on HSC made with OPC & PPC", *The Indian Concrete Journal*, (2006) : 43-48

[7] Xin Luoa, Wei Suna and Sammy Yin Nin Chanb, "Effect of heating and cooling regimes on residual strength and microstructure of normal strength and high-performance concrete", *Cement and Concrete Research* 30 (2000): 379-383.

[8] N.Anand, G. Prince ArulrajInternational Journal of advances in science Engineering and Technology ISSN: 2321-9009 volume 2, Issue-4, oct-2014. "Experimental Investigation on Mechanical properties of self-compacting concrete under elevated temperature".

IOP Conf. Series: Materials Science and Engineering **410** (2018) 012007 doi:10.1088/1757-899X/410/1/012007

[9] W.Brostow and G. Martinez-Barrera Mechanical properties of self-compacting concrete reinforced with polypropylene fibers O. Gencel, C. Ozel,.
[10] Okamura H, Ouchi M, Self-Compacting Concrete Development, Present and Future, In: Å. Skarendahl, Ö. Petersson (eds.), Proceeding of 1st International Rilem Symposium on Self Compacting Concrete; Stockholm, Sweden, 1999, p. 804. e-ISBN: 2912143721.
[11] Saeed Ahmed Imran A Bukhari,Javed Iqbal Siddiqui Shahzad Ali Qureshi A Study On Properties Of Polypropylene Fibre Reinforced Concrete', University Of Engineering & Technology Taxila, Pakistan.

IOP Conf. Series: Materials Science and Engineering **410** (2018) 012008 doi:10.1088/1757-899X/410/1/012008

Synthesis, characterization of TiO_2 nano particles for enhancement of electron transport application in DSSC with Cu-BPCA Dye

Jagannath Panda[1], Udai P. Singh[2], Rojalin Sahu[1]*

[1]*School of Applied Sciences, KIIT Deemed to be University, Bhubaneswar-751024, Odisha, India*

[2]*School of Electronics Engineering, KIIT Deemed to be University, Bhubaneswar-751024, Odisha, India*

ABSTRACT: Dye-sensitized solar cells [DSSCs] have attracted extensive attention due to their potential low cost and high energy efficiency, rendering them one of the most promising system for solar-to-energy conversion. The DSSC efficiency was enhanced by intermixing with the use of TiO_2 nanoparticles which provides high surface area for accommodating the light-absorbing sensitizer and also the stable conductor for photo generated electrons. In hydrothermal method, the TiO_2 nanoparticles synthesis depends on temperature. TiO_2 nano particles diameter depends on different autoclaving temperature.TiO_2 nanoparticles have been coated on ITO glasses by screen printing method. In this work, we have synthesized TiO_2 nano particles which can provide a fast way for electron transport and reduced trapping of photo injected electrons during the path of back contact. The DSSCs were fabricated using the ruthenium dye and electrolyte (I_3/I_3^-). The crystalline structure of TiO_2 has been characterized by DLS, X-ray diffraction, SEM and TEM. The absorption spectra measured by using UV-Vis spectrometer. The IR spectrum has been recorded to know the peaks of Ti-O-Ti in powder sample. It has been found that the efficiency of DSSC was highly affected by the properties of nano particles.

Keywords: Titanium Dioxide (TiO_2), Nanoparticles, Dye, DSSC

1. Introduction

Titanium-Di-Oxide is one of the most attractive material in the field of nanoscience also in nano technology because of the potential application of TiO_2 in different processes and it has its extraordinary properties like low cost, high stability, biocompatibility, non-toxicity, high chemical inertness, high refractive index ,wide band gap etc.[1,2,3] It has been already used in the application of efficient photo catalyst for water purification [4] and degradation of dyes[5], catalyst support[6], chemical sensor[7,8], optical device[9,10], electrochemistry, microelectronics and dye sensitized solar cell[11] etc. The important characteristic which affect the performance of TiO_2 are its particle size, crystallinity and morphology etc [12]. It generally shows three crystalline phases as anatase (Tetragonal), rutile(Tetragonal) or brookite (Orthorhombic) and

IOP Conf. Series: Materials Science and Engineering **410** (2018) 012008 doi:10.1088/1757-899X/410/1/012008

due to these it is used for optical device[13]. The photo catalytic activity increased by optimizing the size using various synthesis processes like sol-gel[14,15], solvothermal[16], co-precipitation, chemical vapour synthesis[17], reverse micelle method, liquid phase deposition etc. This above processes are liquid phase process and synthesis can be carried by using liquid titanium like titanium tetrachloride, titanium alkoxide as a precursor. These method of preparations are very challenging because the preparation of the required concertation of the precursor and the molar and P^H ratio of water with liquid titanium should optimized. Also the use of organic precursor leads to the contaminant of the reaction environment. To avoid the high cost of liquid titanium salt and above reaction conditions, we have used the TiO_2 micro particle powder as precursor for the synthesis of the nano sized particles by hydrothermal process.

2. Experimental section

All chemical used in the experiment were analytical grade. Titanium dioxide (TiO_2) from (MERCK) has been used for the synthesis.

2.1 Synthesis of TiO_2 nano particles

The TiO_2 micro particles powder was mixed with 10M NaOH and treated hydrothermally at 150°C in an autoclave for 24 hours. After hydrothermal reaction the sample was washed by 0.1M HCl solution to make the sample neutral of P^H 7. Then the white solid was dried at 80°C for 24 hours. After drying the obtained powder was heated at 500°C in a hot plate of static air for 2 hours to produce the TiO_2 nano particles.

2.2 Preparation of TiO_2 slurry and Cu-BPCA dye solution.

The TiO_2 slurry was prepared by mixing of 6 ml of ethanol, 3.6 gm of TiO_2 nano particles (NPs) and 1 ml of HCl and then stirred for 2 hour with the addition of 0.5-1 ml of deionised water. The dye solution was prepared by mixing Cu-BPCA (dye) with methanol and Poly vinyl alcohol (PVA) for its good adhesive purpose as surfactant.

2.3 Preparation of DSSC device

a) Deposition of TiO_2 film

We have identified the conducting side of a tin oxide-coated piece of glass by using a multimeter to measure the resistance. The conducting side have a resistance of 20-30 ohms. With the conducting side up, taped the glass on three sides using one thickness of tape. Wiped off any fingerprints or oils using a tissue wet with ethanol. Added some of the titanium dioxide suspension and quickly spreaded using a glass rod. The tape served as a 40-50 micrometer spacer to control the thickness of the titanium dioxide layer. Heated the glass on a hotplate in a hood for 10-20 minutes.

b) Staining TiO_2 with the dye

Immersed the coated glass piece in a solution of Cu-BPCA dye. The white TiO_2 has been changed color as the dye was absorbed and complexed to the Ti(IV) ensuring the dye bound to the TiO_2 film. When visible light was absorbed by dye, excited electrons are formed. This in turn transferred an electron to the TiO_2 which leads to the generation of electricity in the DSSC.

c) Carbon coating on the counter electrode

To prepare the counter electrode, the second piece of tin oxide glass, putting the conducting side down, through a candle flame to coat the conducting side with carbon (soot). In between the two layers, Iodine solution was kept as an electrolyte solution.

2.4 **Material characterisation**

The characterisation of the synthesized TiO_2 nano particles were carried out by using PXRD (Shimzdu, Japan), DLS (Malvern, Germany),UV-Visible spectrophotometer (Shimzdu, Japan) and FT-IR (Thermoscientific, NICOLET iS5 ID3). The photo current production has been measured by I-V measurement system by using solar simulator instrument.

3. **Result and discussion**

3.1 **PXRD**

X-ray diffractometer (Shimatzu-6100) with CuKα, $\lambda = 1.54$ Å wavelength has been used to record the PXRD of the samples. Θ-2θ scans were carried out from 10^0 to 80^0 with scan speed of 2^0/min with incidence angle of 10^0 and step size of 0.02^0. In the figure-1, the PXRD pattern of the TiO_2 nano particles has been shown. It has been confirmed the image that it is having the anatase phase with good crystalline nature. The average crystalline size of the TiO_2 nano particles can be found to be 6-100 nm[18]. The particles are crystalline in nature with 2θ peaks at 2θ = 24.62 (101), 2θ = 28.90 (110), and 2θ = 48.42 (200) . The peaks in the PXRD patterns can be indexed as anatase phase of TiO_2 and the diffraction data was in good agreement with JCPDS files # 751816.

Crystallite size was obtained by Debye-Scherrer's formula given by equation

$$D=K\lambda/(\beta\cos\theta)$$

Where D is the crystal size; λ is the wavelength of the X-ray radiation (λ=0.15406 nm) for CuKa; K is usually taken as 0.89; and ß is the line width at half-maximum height. The crystallite size obtained using this formula is 2.3 nm [19] .

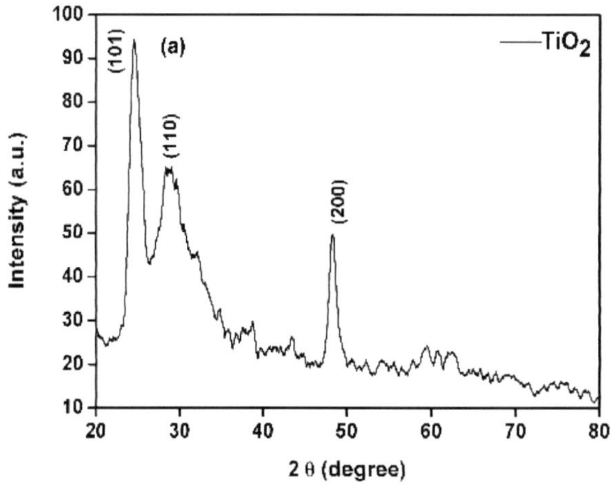

Figure-1: (a) PXRD plot of TiO_2 nanoparticle

3.2 **UV-Visible spectroscopy**

The band gap of TiO_2 nano particles were calculated from the UV-Vis spectra which is shown in Fig.1(b).

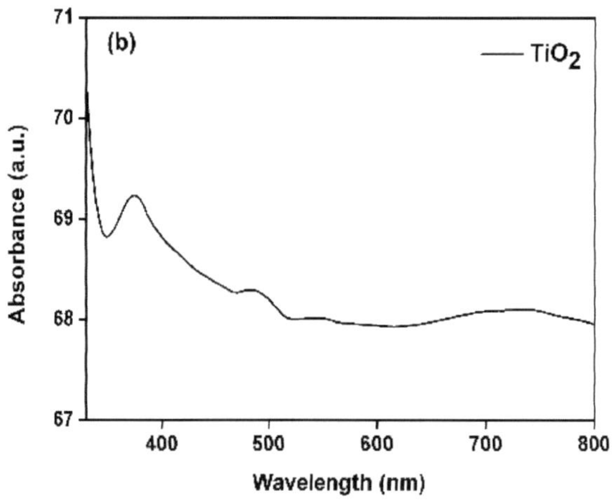

Figure-1: (b) UV-Visible spectra of TiO_2 nano particles

We got the absorbance at 350 nm and the band gap is found as 3.5 ev. It has been concluded from the UV absorbance spectra that TiO_2 reacts with intense UV light and instantly act as a photo catalyst through its hydroxyl radical causing breaking of strong covalent bonds [20].

3.3 **FT-IR**

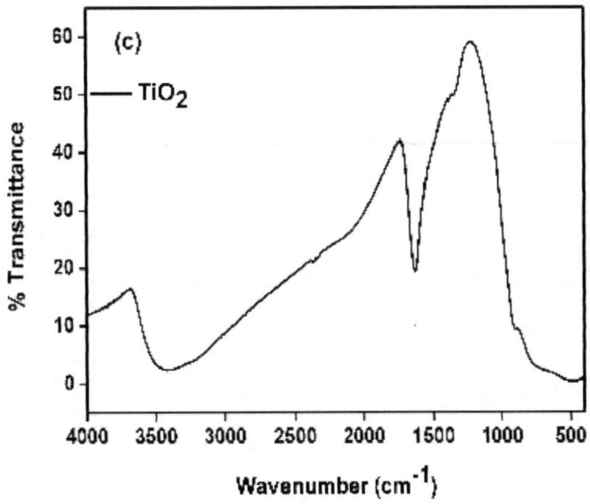

Figure-1: (c) FT-IR (d) DLS of TiO_2 nano particles.

The absorption spectra in the infrared region was acquired by using Thermo Scientific Nicolet iS5 spectrophotometer equipped with an ATR accessory with a resolution of 4 cm^{-1} in the range of 400-4000 cm^{-1}. The peaks at 3400 and 1650 cm^{-1} in the spectra are due to the stretching and bending vibration of the –OH group. A small peak at 1450 cm^{-1} shows stretching vibration of Ti-O-Ti [21].

3.4 **Dynamic Light Scattering**

TiO_2 nano particles were characterised for its distribution of sizes in 0.1 % w/v TiO_2 NPs sample dispersed in 10 mM NaCl by using Zeta Sizer Nano ZS at 25°C. The particle size distribution has been obtained from Malvern, Germany dynamic light scattering instrument. The z-average diameter was measured 421.1 nm. Some of the particles are in the size of 108.2 nm which was much smaller than the expected sizes [22].

			Size (d.nm):	% Intensity:	St Dev (d.n...
Z-Average (d.nm):	421.1	Peak 1:	958.7	83.0	377.4
PdI:	0.611	Peak 2:	108.2	17.0	28.86
Intercept:	0.793	Peak 3:	0.000	0.0	0.000
Result quality :	Refer to quality report				

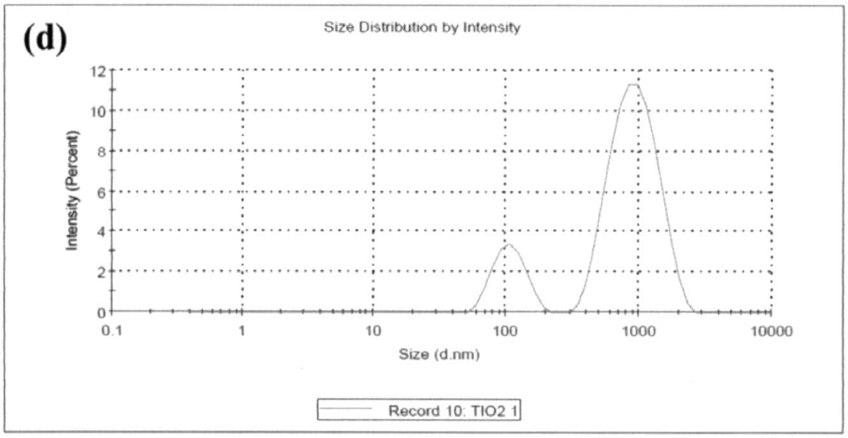

Figure-1: (d) DLS of TiO_2 nano particles.

3.5 **Photoluminescence**

Photoluminescence specta of TiO_2 nano particles were recorded in an excition frquency of 479 nm and at room temperture.

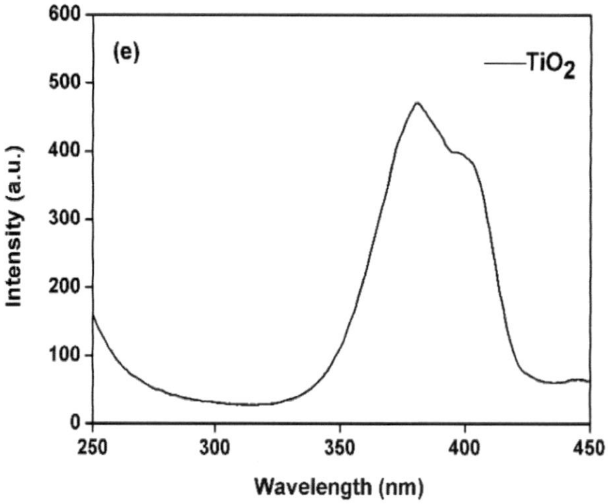

Figure-1: (e) Photo Luminescence (PL) of TiO_2 nano particles.

The photoluminscence spectra in anatase TiO_2 form has been shown in Fig.1(e). which is originated due to self traped excitions, surface state and oxygen vacencies[23]. The oxygen vaccancy peak normaly shows at a range of 391 nm to 409 nm.In our sample it showed at these ranges which indicated the photoluminescence emissions happened due to the vaccancy of oxygen in the TiO_2 nano particles [24].

3.6 **Transmission Electron Microscopy & Scanning Electron Microscopy**

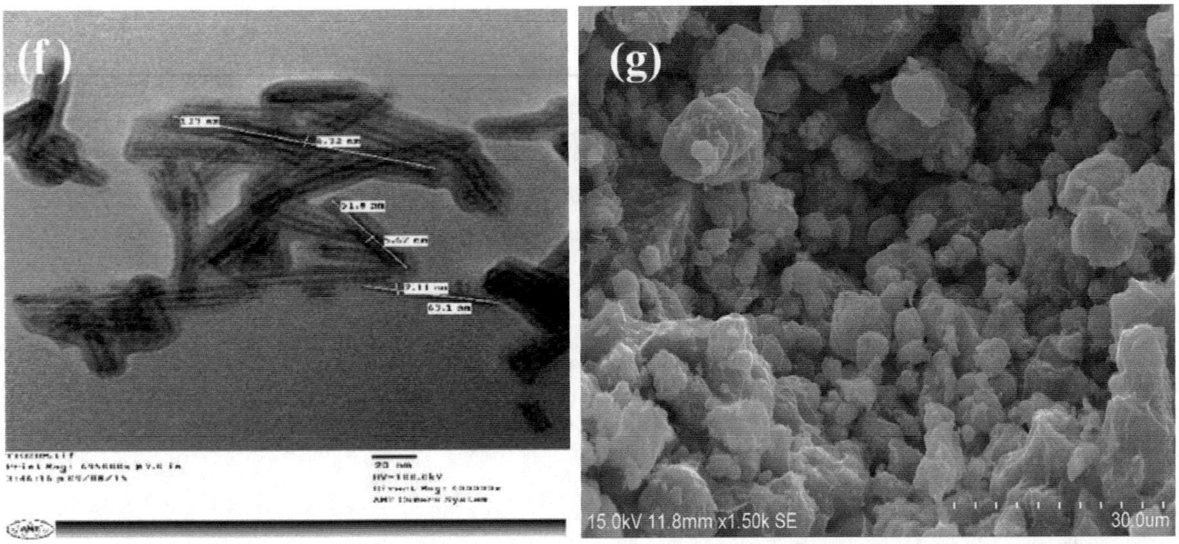

Figure-1: (f) TEM (g) SEM of TiO_2 nano particles

TEM images of hydrothermal derived nanoparticles are shown in Fig. a. Clear rod shape structures can be seen in the Fig. a having diameter 6 ~ 100 nm. Selected area diffraction is shown in inset of Fig. Which clearly indicates that the TiO_2 nanoparticles are highly crystalline in nature As can be seen from the TEM image that the average particle size is ~ 20 nm, which is in agreement with the crystallite, size obtained from XRD[25]. The SEM images in fig-b shows particles with great aggregation. The shape of the particle is not uniform and it looks like spherical in shape.[3]

3.7 Band gap of TiO_2 nanoparticles

The band gap is found as 3.5 ev. It has been concluded from the UV absorbance spectra that TiO_2 reacts with intense UV light and instantly act as a photo catalyst through its hydroxyl radical causing breaking of strong covalent bonds [20].

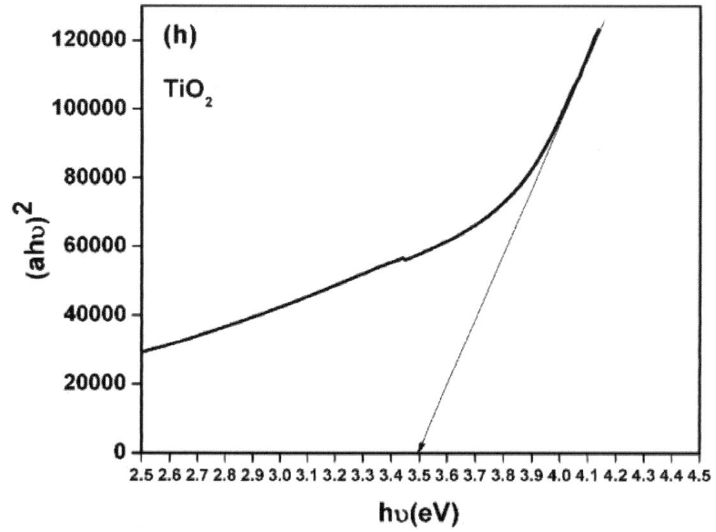

Figure-1: (h) Band gap of TiO_2 nano particles.

3.8 **I-V Characterization**

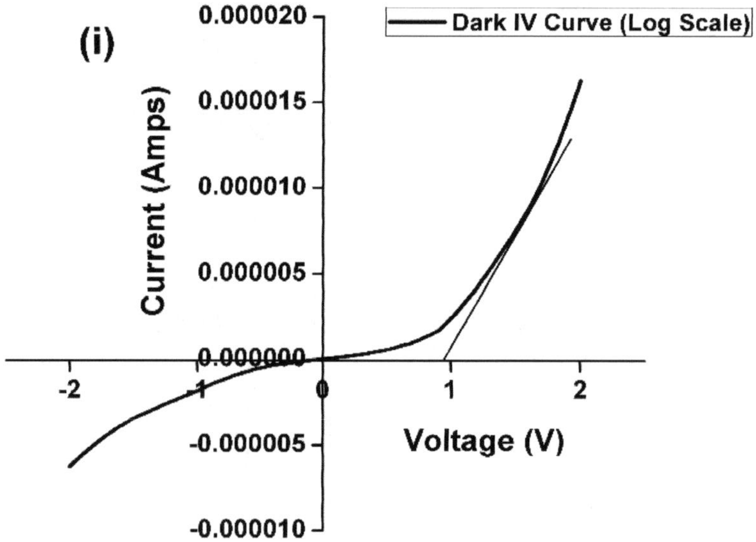

Figure-1: (i) I-V plot of TiO_2 nano particles.

The current - voltage characteristic were performed using the instrument Keithley model 2611 source measuring unit. We have calculated the I-V Characteristics in dark condition.

From the Fig. 1(i), it has been seen that an ideal diode characteristic has been shown by the device. The Knee Voltage (V_k) is found to be approximately 0.9 V [26].

4. Conclusion:

TiO_2 nanoparticles have been successfully synthesized adopting a clean and economical synthesis protocol, i.e, hydrothermal method. The formation of the TiO_2 nanoparticles has been confirmed by powder X-ray diffraction (PXRD). The size and morphology of the samples were characterized using scanning and transmission electron microscopy (SEM and TEM). The irregular spherical shaped particles were confirmed through the SEM analysis. The transmission electron microscopic analysis confirms the synthesized TiO_2 nanoparticles with the particle sizes in the range of 6-20 nm having nano rod shapes. The DSSC device was showing an ideal diode characteristic which is the basic requirement of a solar cell.

Acknowledgment
The authors acknowledge the KIIT Deemed to be University, Bhubaneswar for the financial support.

Reference:

1. Khan, M.M.; Ansari, S.A.; Pradhan, D.; Ansari, M.O.; Han, H.D.; Lee, J.; Cho, M.H.; J. Mater. Chem. A, 2014,2, 637-644
2. Nolan, M.P. N.T.; Pillai, S.C.; Seery, M.K.; Falaras, P.; Kontos, A.G.; Dunlop, P.S.M., Hamilton, J.W.J.; Byrne, J.A.; Sheaf, K.; Entezari, M. H., Dionysiou, D. D.; Appl. Catal., B, 2012, 125, 331–349.
3. Kavitha, M.; Gopinathan, C.; and Pandi, P.; International Journal of Advancements in Research & Technology,2013,2, 4.
4. Ocwelwang, R.A.; Tichagwa.L.; International Journal of Advanced Research in Chemical Science (IJARCS),2014,1, 2,28-37.
5. Khataee, A.R.; Zahraa, P.O.; J. Hazard. Mater., 2009,168,1, 451–457.
6. Ba-Abbad, M.M.; Kadhum, A.A.H.; Mohamad, A.B.; Takriff, M.S.; Sopian, K.; Int. J. Electrochem. Sci., 2012, 7, 4871 – 4888.
7. Kulwicki, B.M.; J.Phys. Chem. Solids, 1984, 45,1015–1031.
8. Micheli, A.L.; Am. Ceram. Soc. Bull. 1984, 63,694–698.
9. Pal, K.; Majumder, T. P., Neogy, C.; Debnath, S. C., J. Mol. Struct.,2012, 1016, 30–38.

10. Allen N. S.; Edge M.; Verran J.; Stratton J.; Maltby J.; Bygott C.; Polym. Degrad. Stab., 2008, 93, 1632-1446.
11. Wang Y.; Wu J. H.; Lan Z.; Xiao, Y.; Li, Q.; Peng, F.; Lin, J.; Huang, M.; Chinese Sci Bull, 2011, 56,2649-2653,
12. Monticone, S.; Tufeu, R.; Kanaev, A.V.; Scolan, E.; Sanchez, C.; Appl. Surf. Sci. 2000,162,565–570.
13. Thamaphat, K.; Limsuwan, P.; Ngotawornchai, B.; Nat. Sci., 2008, 42, 357 – 361.
14. Hema, M.; Arasi, A.Y.; Lselvi, P. T.; Anbarasan, R.; Chem Sci Trans., 2013, 2, 239-245.
15. Lee, C.H.; Rhee, S.W.; Choi, H.W.; Nanoscale Res. Lett. 2012, 7,48
16. Marco,L.D.; Manca,M.; , Giannuzzi,R.; Malara ,F.; Melcarne,G.; Ciccarella ,G.; Zam, I.; Cingolani,R.; Gigli,G.; J. Phys. Chem. *C*, 2010, 114, 4228–4236
17. Djenadic, R.; and Winterer, M.; Nano Science and Technology, 2012, 49-76.
18. Sagadevan, S.; Am. J. Nanosci. Nanotechnol. 2013, 1,27-30
19. Vijayalakshmi, R.; Rajendran, V.; Arch. Appl. Sci. Res., 2012, 4, 1183-1190.
20. Jwo, C.S.; Tien D.C.; Teng T.P.; Chang H.; Tsung T.T.; Liao C.Y.; Lin, C. H.; Rev. Adv. Mater. Sci. 2005,10, 283-288.
21. Thangavelu, K.; Annamalai, R.; Arulnandhi, D.; IJETAE, 2013,3.
22. Piskin, S.; Palantöken, A.; Yilmaz M.S.; International Conference on Emerging Trends in Engineering and Technology (ICETET'2013),2013,7-8,Patong Beach, Phuket (Thailand)
23. Saraf, L. V.; Patil, S. I., Ogale, S. B.; Int. J. Mod. Phys. B, 1998,12, 2635-2647.
24. Ho, C.Y.; Lin, J. K.; Wang, H.W.; Solid State Commun., 1993, 87, 847-850.
25. Razali, M.H.; M. N, A.F.; Sreekantan, S.; Mohamed,A.R.; Adv. Mater. Res.,2013, 772, 365-370.
26. Merlino, R. L.; Am. J. Phys. 2007, 75, 1078-1085.

IOP Conf. Series: Materials Science and Engineering **410** (2018) 012009 doi:10.1088/1757-899X/410/1/012009

Effect of fullerene content on the thermal, microstructure, and electrokinetic properties of fullerene/poly(vinyl pyrrolidone) nanofluids and nanocomposites

M. Behera

Silicon Institute of Technology, Bhubaneswar-751 024, India
email: mano.silicon@gmail.com

Abstract

In this study I report the thermal stability, microstructures, dynamic light scattering (DLS) and electro-kinetic effect in the fullerene (C_{60}):poly(vinyl pyrrolidone) PVP nanofluids (NFs) and nanocomposite films. An improvement in the degradation temperature of C_{60}:PVP nanocomposite reported here is ascribed to free radical scavenging ability of C_{60}. An analysis of DLS data in terms of polydispersity index (PDI) reveals a well dispersed structure of C_{60} assemblies with PVP in liquid medium. The typical electro-kinetic parameters such as zetapotential (ξ) and surface conductivity (σ_{sc}) in a NF evaluates the magnitude of electrostatic repulsion between C_{60} particles and such measurement tender a way to boost stability of C_{60} nanoparticles in presence of PVP molecules in a particular carrier. Microstructures in various films were found to vary with increasing in the C_{60} content and decreasing in the relative PVP content in the nanocomposites. Highest thermal degradation temperature (~ 422 ^{0}C) was observed for the sample consisting of 13.9 μM C_{60} owing to uniform distribution of C_{60} in the PVP matrix with a low PDI (0.272).

1. Introduction

Thermal degradation of raw macromolecular compounds at the early stage is a major setback to polymer industry [1,2]. In recent years many studies have been focused on improving their thermal stability by introducing nanofillers into the host matrix by both physical and chemical processes [3-8]. Miniscule fillers mostly used in improving the thermal properties of polymers could be clay, metal nanoparticles (NPs), metal oxides, carbon nanotube, and fullerene (C_{60}). Although many mechanisms have been proposed to explain enhancement in thermal properties of polymer nanocomposites upon introduction of NPs, the radical trapping of free radicals believe to be suite the most. Among the various classes of nanofillers, C_{60} is one of the most widely noticed filling agents because of its high reactivity towards free radicals [5-12] In an article [12], it reported that uniform dispersion of Pd NPs in the polymer matrix increases the thermal decomposition temperature of polypropylene (PP), polystryrene (PS) and polymethyl methacrylate (PMMA).

A three dimensional free radical scavenger like C_{60} has been reported by authors that it improves the thermal stability of various polymers [5-12]. In an article, Zhao et al. [5] reported that the resistance to the thermal degradation of high density polyethylene, PP, PMMA, and bisphenol-A polycarbonate were improved a lot with the addition of C_{60}. Their results showed that the influences of C_{60} on the resistance to the thermal degradation of different polymers were dependent on radical trapping thermal degradation mechanism. The influence of C_{60} on the thermal degradation of PMMA and polyvinyl chloride (PVC) is connected with its interruption of a chain radical reaction of the thermal degradation of PVC and PMMA. Gaspar et al. [6] have compared with pure PMMA, both C_{60} and PCBM increase substantially the thermal degradation temperature of the corresponding composites, both in nitrogen and air atmospheres. It is reported by Katiyar et al. [7] that incorporation of C_{60} in the PMMA increases the both onset temperature of polymer degradation/decomposition and temperature of maximum weight loss only because of radical scavenging effect of C_{60}. The impact of two different fullerenes (C_{60} and a functionalized C_{60}) on the thermal stability of the corresponding polystyrene (PS)-fullerene composites was studied by Fernandes et al. [8] using dynamic thermo gravimetric analysis. According to them the improved thermo-oxidative stability of the functionalized-fullerene composites compared to the C_{60} composites is due to the fact that former disperses better in a PS matrix than C_{60}.The results showed that the influence of C_{60} to resist early thermal degradation of different polymers were dependent on their degradation mechanism. The resistance to the thermal degradation of many polymers such as HDPE, PP, and PMMA were improved with the addition of C_{60} indicated that the radical trapping route played a dominant role. Kukobo et al. [9] have reported that the thermo-oxidative stability of PMMA/C_{60} nanocomposite with optimal loading (0.4–0.8 wt% C_{60}) exhibited larger thermal stability over PMMA. They attributed this enhancement to the radical-scavenging ability of π-conjugative fullerenes. In another article, Song et al. [10] studied the influence of C_{60} on the thermal properties of PP and found that C_{60} improved the thermal properties of PP by trapping free radicals and in situ forming a gelled-ball crosslink network. Zuev et al. [11] studied that the influence of C_{60} on the thermal behavior and thermo-degradation of acrylic polymers. The results showed that the decomposition process of acrylic polymers changed from a radical pathway to a non-radical pathway with the addition of C_{60}.

In nano-systesms, since the strong van der Waals interactions between small structures always favor aggregation and afterward sedimentation, developing a stable nanofluid (NF) is not an easy task for the chemists. In general, stability in a NF comes as a result of interplay of the heat (kinetic) energy of the flocculates, attractive forces between the flocculates, and repulsive forces between the flocculates and the liquid carrier. It is reported that, particles with zetapotential (ξ) $\geq$ ($\pm$) 30 mV are normally considered as stable in a suspension [13,14]. The electrostatic stabilization of small structures in a suspension is successfully described by the DLVO theory, named after the work of Derjaguin, Landau, Verwey, and Overbeek [13,14]. In an experiment, Rahme et al. [15] found that the ξ-value in a NF also depends on the host structure. A ξ-value had been decreased from (-) 35 mV to (-) 1 mV in a gold reinforced polyethylene glycol (PEG) NF in a nonlinear path when PEG molecular weight was increased from 2,100 to 51,400 g mol^{-1}. They remarked that such a decrease in the ξ-value with an increased molecular weight of a liquid carrier is not surprising as zeta potential is measured at the surface plane of the hydrodynamic sphere of diameter and the surface plane of the hydrodynamic sphere is far from the surface of flocculates in a stable NF.

The dynamic light scattering (DLS) is a technique which is capable of measuring average size of flocculates like C_{60} clusters in terms of (L_{hd}) and width of distribution as polydispersity index (PDI) in a NF [13,16.17]. In DLS, the speed at which the nanoparticles (NPs) are diffusing in a NF due to the Brownian motion over a liquid carrier is measured very precisely. As the NPs in a NF are under constant, but random Brownian motion, they cause the intensity of scattered light from them to fluctuate as a function of time. In general, the small particles cause the intensity to fluctuate more rapidly than effectively larger ones. So, the L_{hd}-value of C_{60} in a NF can be obtained from the temporal autocorrelation function relative to the diffused light intensity. From the literatures [13,14], the hydrodynamic diameter L_{hd}-value is obtained from a field correlation function $g^1(\tau)$ which is an exponential decaying function of a correlator time delay τ;

$$g^1(\tau) = A[1 + \beta \exp(-2\Gamma\tau)], \quad (1)$$

where A is a base line, β is an intercept of the correlation function, which represents the degree of spatial coherence of the scattered light over the detector, and it is determined by the ratio of the detector area to the area of one speckle ($0 < \beta < 1$), while $\Gamma = Dq^2$ describes the particle relaxation with an autocorrelation time τ (= $1/\Gamma$) and a translational diffusion coefficient D. The τ-value gives information about a dynamic fluid behavior. The scattering wavevector, $q = (4\pi\mu/\lambda) \sin(\theta/2)$, is determined by the refractive index μ of the medium, wavelength of laser λ and the scattering angle θ. For model spherical particles, the D_0 value (D =

D_0 at infinite dilution of a fluid) relates to the effective L_{hd} value in a well-known Stokes-Einstein relation [16,17]:

$$D_0 = kT/3\pi\eta L_{hd}, \quad (2)$$

which yields the L_{hd}-value at temperature T after putting the viscosity (η) of the solvent and the Boltzmann constant $k = 1.38 \times 10^{-23}$ JK^{-1}.

Alagrova et al. [18] studied the variation of $g^1(\tau)$ against decay time for a colloidal dispersion of C_{60} in a toluene/acetonitrile mixture solution in absence of stabilizing agent. As they have reported, a PDI of 0.1 appears from a second-order cumulant fit analysis. In another experiment, Chaudhari et al. [19] studied the variation of electric field correlation function of 50 mM sodium dodecyl sulfate surfactant (SDS) at different anilium nitrate (AN) concentrations. They reported that relaxation time of $g^1(\tau)$ increases with increasing AN concentration due to an increase in effective size of the micelles and L_{hd}-value of SDS micelle has increased almost linearly with AN concentration in the micellar solution. There is a report by Mountrichas et al. [20] that both the PDI indices and average hydrodynamic diameters (L_{hd}) are increased upon increasing the C_{60}-content from 1–20 wt% in amphiphilic copolymer solubilized C_{60} in water. In C_{60}-polystyrene NFs in benzene, it is reported that the L_{hd}-value of C_{60} NPs decreases from 220 nm to 140 nm when concentration of the polymer has increased from 0.56 M to 2.5 M [21]. In another experiment, Rahme et al. [15) reported that L_{hd}-value of Au NPs has increased in a nonlinear way from 10 to 90 nm by increasing molecular weight of the polymer from 2,100 to 51,400 g mol^{-1} in grafted Au-PEG NFs in water due to a decrease in grafting density from 3.93 nm^2 to 0.32 nm^2. In Au-poly(acrylic acid) (PAA) NFs, it is reported that the L_{hd}-value of Au-NPs has decreased from 6.5 nm to 1.4 nm when the PAA content has increased from 0.006 mM to 6.0 mM in water [22]. In Au:PVP NFs in water, Mishra et al. [23] have reported a shift in the correlograms over larger L_{hd} values on increasing the Au-content from 2.54 to 50.76 mM, supporting a fact that a solid inclusion combine to PVP in joint assemblies. In an article on Au-PVA NFs, it is reported that with increase in the Au-content, the correlogram had shifted to right along the X-axis and that causes in delaying the time τ [24]. A comparison of all these above results implies that the structural parameters in a NF depends on several factors namely (i) initial size of the solid particles, (ii) interaction of the solid particles with the polymer additives if any used so that they form joint assemblies, and (iii) average concentration of the NF made in a particular liquid carrier.

In this article I discussed the thermal stability, the variation of degradation temperature, L_{hd}, PDI, ξ-value and surface conductivity (σ_{sc}) of PVP/C_{60} clusters present in the C_{60}:PVP NFs/nanocomposites of selective compositions with C_{60} content.

2.0 Synthesis of NFs and experimental measurements

99.9% purity C_{60} powder and K-25 grade PVP of average molecular weight of 28 kDa were purchased from Alfa Aesar and Aldrich chemicals, respectively. Toluene and n-butanol non-aqueous solvents of analytical grade were used to develop NFs. Getting information on the synthesis of C_{60}:PVP NFs in n-butanol, published article [25] may be referred. Thermo-Gravimetric analysis (TGA) was carried out using a TA Q500 thermo-balance. The instrument was calibrated with respect to indium and aluminum standards. Samples (~5 mg) were placed in alumina crucibles and dynamic tests were performed by heating the samples from 40 ^{0}C to 600 ^{0}C using a heating ramp of 10 ^{0}C min^{-1}, under nitrogen at 50 mL min^{-1}.The DLS experiments were carried out for the parent PVP solution and derived C_{60}:PVP NFs in *n*-butanol consisting of selective C_{60} content with the help of Malvern Nano Zetasizer Nano ZS instrument. An incident light beam was generated from a vertically polarized He-Ne laser beam of wavelength $\lambda = 633.2$ nm (of 4 mW power) fixed at one arm of a Goniometer. Before measuring the data, a NF was filtered through a 0.22 μm Teflon filter paper and loaded into a 3 mL quartz cell. The same instrument was also used to measure the electro-kinetic parameters of various NFs.

3.0 Results and discussion

3.1 DLS bands of C_{60}:PVP NFs

IOP Conf. Series: Materials Science and Engineering **410** (2018) 012009 doi:10.1088/1757-899X/410/1/012009

Figure 1 depicts the normalized $g^1(\tau)$ values measured for C_{60}:PVP NFs comprising (a) 0, (b) 4.63, (c) 9.26, (d) 13.9, (e) 18.52, and (f) 23.15 μM C_{60} along with 40.0 g/L PVP in *n*-butanol. A linear Γ dependence has been shown upon the q^2-value in describing $g^1(\tau)$ of a single (translational) diffusive relaxation mode. As it can be seen from the plot in Figure 1, with increase in the C_{60}- content, the correlogram hardly shifts along the X-axis. This reveals that a C_{60} doping barely influence the L_{hd}-value in a NF, and consequently has a little effect on the delay time τ. In general, if the C_{60} clusters are effectively large in size, the signal will change slowly and the correlation will persevere for a long time. If the particles are effectively small and moving rather rapidly, then correlation will reduce more quickly, and a correlogram will make a left-shift along the time-axis.

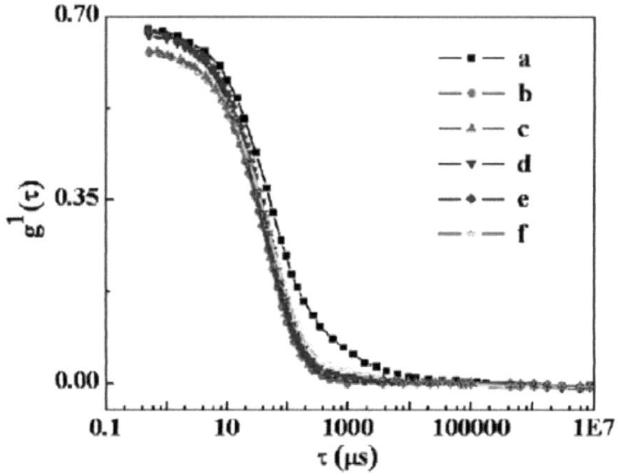

Figure 1. Variations of normalize $g^1(\tau)$ values against the time scale of measurement in C_{60}:PVP NFs in *n*-butanol; (a) 0, (b) 4.63, (c) 9.26, (d) 13.9, (e) 18.52, and (f) 23.15 μM C_{60}.

Further I studied DLS bands in the C_{60}:PVP NFs prepared with the different C_{60} dosages (a) 0, (b) 4.63, (c) 9.26, (d) 13.9, (e) 18.52, and (f) 23.15 μM in 40.0 g/L PVP in *n*-butanol. The experimental data could be collected over five times diluted samples which had an average 6.8 pH.

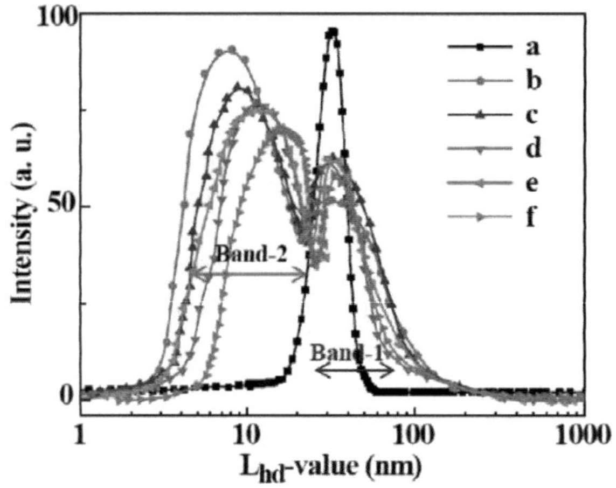

Figure 2. Distributions of L_{hd}-values in C_{60}:PVP NFs consisting of (a) 0, (b) 4.63, (c) 9.26, (d) 13.9, (e) 18.52, and (f) 23.15 μM C_{60} in 40.0 g/L PVP in *n*-butanol.

As it is can be seen from the DLS spectra portrayed in Figure 2, except the virgin sample (a) without having any C_{60} doping, all the C_{60}:PVP NFs exhibit two distinct DLS bands, indicating two types of size distributions of the flocculates. In sample (a) of only PVP molecules dispersed in *n*-butanol exhibits a single DLS band {full width at half maximum (fwhm) ~15.8 nm} with an average L_{hd}-value 30 nm and a PDI 0.298. It is observed that when only a small dosage of 4.63 μM C_{60} molecules is added to this sample, two well-separated bands split-up in two effective L_{hd}-values 32.0 and 8.1 nm. An average L_{hd}-value 9.2 nm is estimated from the total band profile, with PDI 0.35. The first Band-1 of 32.0 nm, which is reasonably broad (fwhm ~48.7 nm), attributes to PVP molecules dispersed in *n*-butanol. The Band-2 with fwhm ~15.9 nm arises in PVP-capped C_{60} particles. It is also examined from the plot that with an increase in the C_{60}-value, the average L_{hd}-value of Band-1 in the PVP molecules is almost unchanged whereas that has increased nonlinearly to as large as 17.0 nm in Band-2.

From the results in Table 1 for the various C_{60}:PVP NFs, a pretty small average L_{hd}-value 9.2 nm is found for a C_{60}:PVP NF, which consists of a reasonably small doping of 4.63 μM C_{60}, and that has grown-up significantly to 17.0 nm, i.e., nearly twice the above value upon increasing the doping dosage to 23.15 μM C_{60}. These results explicitly confer formation of relatively larger flocculates between the two kinds of the entities of PVP and C_{60} as a result that the C_{60} clusters and grows dispersed through PVP molecules which control the C_{60}:PVP clustering in a dispersed sample of a NF. Evidently, a PVP polymer solution not only acts as a C_{60} solubilizer in *n*-butanol, but also extends an excellent surface encapsulant so that it retards C_{60}:PVP clustering in small assemblies. Technically, this is highly required to produce a stable NF with functionalized properties.

Table 1: The values of ξ, σ_{sc}, L_{hd}, and PDI determined from zetapotential and DLS for PVP solution and synthesized C_{60}:PVP NFs in *n*-butanol

C_{60} content (μM)	ξ-value (mV)	σ_{sc},-value (mS/cm)	L_{hd}-value (nm)	PDI-value
0	(-) 5.6	0.043	30.0	0.298
4.63	(-) 12.4	0.065	9.2	0.350
9.26	(-) 17.0	0.081	10.0	0.388
13.90	(-) 22.0	0.091	13.0	0.272
18.52	(-) 10.8	0.062	12.6	0.409
23.15	(-) 9.6	0.051	17.0	0.458

(N.B.: The samples were diluted five times in n-butanol before taking the readings)

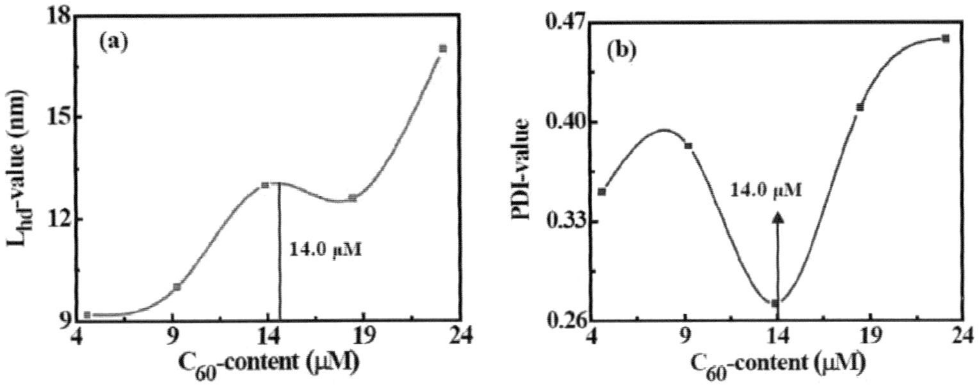

Figure 3. Variations in (a) average L_{hd}-value and (b) PDI-value in the C_{60}:PVP NFs with C_{60}-content prepared in 40.0 g/L PVP in *n*-butanol.

Furthermore, I studied the variations in L_{hd} and PDI values with C_{60}-content in the synthesized C_{60}:PVP NFs in *n*-butanol. As the results are plotted in Figure 3, both these physical parameters trace a nonlinear path with the C_{60} doping of a 40.0 g/L PVP in a solution in *n*-butanol over a wide region of 4–24 μM covered in this experiment. It is evident from the plot in Figure 3a that on increasing the C_{60}-content, the L_{hd}-value displays a peak in a percolation effect near 14.0 μM C_{60}. The C_{60}–content determines how it forms a C_{60}-PVP complex on reaction with exfoliated PVP molecules in a dilute liquid medium. Conversely, the PDI-value Vs C_{60}-content plot made in Figure 3b exhibits a sharp dip near the same value of 14.0 μM C_{60}. As a result, a minumium PDI-value of 0.272 is obtained in sample (d) and it implies that both the C_{60}-PVP and PVP entities are distributed reasonably uniformly in their effective sizes in the two distinct distribution bands in *n*-butanol. This results are differ from the observation of Alagrova et al. [18] that an average L_{hd}-value of C_{60} clusters in colloidal dispersions made in organic solvents such as toluene, benzene, xylene, chloroform, etc. does not changed much on varying the C_{60}-content. Though the reported PDI indices in all these dispersions were reasonably small below a 0.1 value, the L_{hd}-value was rather large lying between 207–240 nm when the C_{60}-content is increased from 0.001 to 0.02 g/L C_{60} (or 1.39 to 27.77 μM) without using a stabilizing agent. These L_{hd}-values are comparatively very large as compared our values estimated in the C_{60}:PVP NFs in *n*-butanol.

3.2 Zetapotential and surface conductivity of C_{60}:PVP NFs

I studied distribution of ξ-values in C_{60}:PVP NFs consisting of (a) 0, (b) 4.63, (c) 9.26, (d) 13.9, (e) 18.52, and (f) 23.15 μM C_{60} in 40.0 g/L PVP in *n*-butanol. As the experimental data could be collected only from reasonably dilute samples, I made samples dilute by 5–10 times from the parent samples in *n*-butanol (average 6.8 pH).

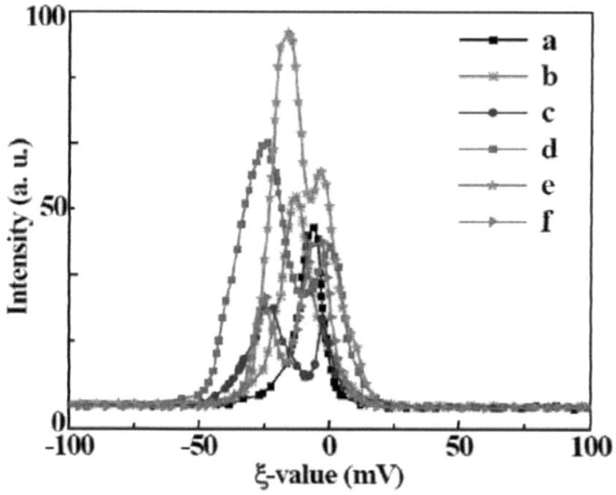

Figure 4. Distribution of ξ-values in C_{60}:PVP NFs consisting of (a) 0, (b) 4.63, (c) 9.26, (d) 13.9, (e) 18.52, and (f) 23.15 μM C_{60} along with 40.0 g/L PVP in *n*-butanol.

As the spectra are shown in Figure 4 for the various samples, a bare PVP solution in sample (a) exhibits a single sharp zeta band at an average ξ-value (-) 5.6 mV with fwhm ~9.1 mV and σ_{sc}-value 0.043 mS/cm in *n*-butanol at 6.8 pH. A negative ξ -value in the bare PVP molecules result from the presence of n-electrons on the >C–N and C=O sites over >N–C=O moieties in PVP molecules. Moreover, a merely 4.63 μM of C_{60} molecules when admixed to sample (a) causes a pretty broader zeta band of an average ξ-value increased to (-) 12.4 mV, with fwhm ~14.4 mV and an enhanced σ_{sc}-value to 0.065 mS/cm. As reported in articles [25-30], this type enhanced values in these two parameters are ascribed to accumulation of n-electrons of >N–C=O moieties from the PVP molecules on the C_{60} surface apparently in the form of a C_{60}:PVP charge transfer complex.

It is also observed from the spectra in Figure 4 that, as the C_{60} doping concentration reaches to a critical value 9.26 μM and above, two types of size distributions of the flocculates emerge in two distinct ξ-bands. Both the bands are prominent in sample (e) which contains 18.52 μM C_{60} molecules. A sample (f) having a still larger C_{60}-content of 23.15 μM displays two well-separated bands apart at (-) 2.2 and (-) 24.0 mV but of relatively poor intensities. The first band is rather stronger and sharing about 60 % integrated intensity. Primarily C_{60} molecules which are encapsulated in PVP molecules in a stable surface layer exhibit this specific band. The other band ascribes to PVP molecules which are merely dispersed via solvent *n*-butanol. An average ξ-value (-) 9.6 mV is estimated from a spectrum of a double band, with σ_{sc} = 0.051 mS/cm.

The values obtained for ξ and σ_{sc} in a bare 40.0 g/L PVP solution and that of reinforced with different C_{60} dosages in the forms of C_{60}:PVP NFs in *n*-butanol are given in Table 1. A comparison of the ξ-values in correlation to L_{hd} values in these samples suggests that a pretty large value, i.e. as much as (-) 22.0 mV, possibly arises as an effectively large number density is set-up of PVP modified C_{60} particles at an optimal doping of 13.9 μM C_{60}, which is the percolation threshold value. Eventually, such a sample shares an effectively large surface-interface area of large number of C_{60} particles so as they adsorb exfoliated PVP molecules in a concomitantly large number over them to give rise an effectively large number of n-electrons localized over the ">N-C=O" moieties of a PVP surface layer. This is what is required to promote an efficient electron transfer of the localized n-electrons from a PVP surface layer to the C_{60} core in a PVP surface modified C_{60} of small particles and/or joint C_{60}:PVP assemblies.

I also studied how the ξ and σ_{sc}-values vary in these C_{60}:PVP NFs with the C_{60}-content in an alcoholic medium. As depicted in Figure 5, both these parameters show two overlapping peaks, one near 5.5 μM C_{60} and other at 13.3 μM C_{60}, which acquires a relatively small viscosity with a reasonably decreased PDI (Table 1) in the C_{60}:PVP NFs.

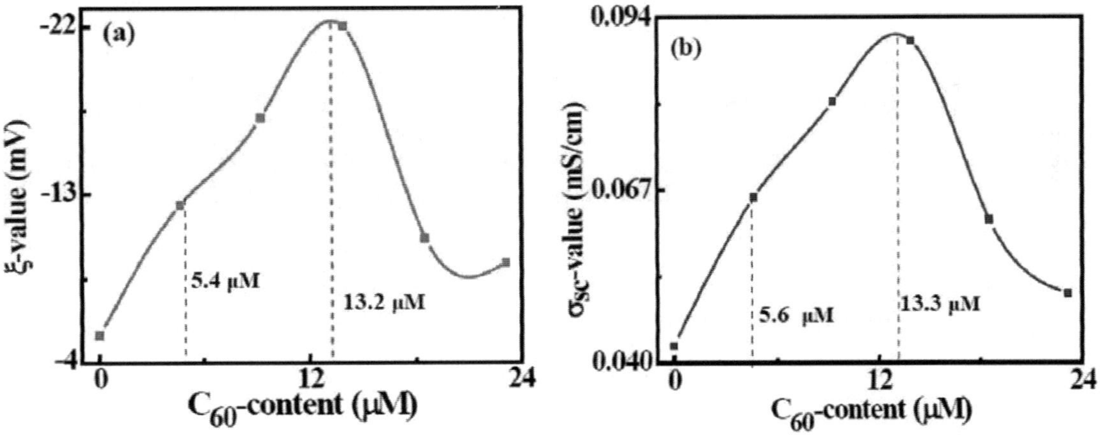

Figure 5. Variations in (a) ξ-value and (b) σ_{sc}-value in a bare 40.0 g/L PVP solution and that grafted with different C_{60} dosages in the forms of C_{60}: PVP NFs in *n*-butanol.

As a result, effectively large ξ and σ_{sc}-values have a correlation to a moderately viscous sample. Such a sample lets a light induced mechanical perturbation propagate through the sample. In the other wards, an exfoliated PVP molecule gets easily adsorbed on a C_{60} nanosurface via n-electron rich C–N and C=O sites in a PVP surface layer in a core-shell structure.

3.3 Thermal analysis of C_{60}:PVP nanocomposites

Figure 6 shows the TGA curves for (a) PVP without C_{60} and C_{60}:PVP nanocomposite films consisting of (b) 4.63, (c) 9.26, (d) 13.9, (e) 18.15, and (f) 23.15 μM C_{60}. From the Figure 6 it was observed that the onset temperatures of thermal degradation of C_{60}:PVP nanocomposite films were higher than that of pure PVP. For C_{60}:PVP nanocomposite films, it increases non-linearly with C_{60} content with a hump near 13.9 μM C_{60} (Figure 7a). The enhancement in the thermal stability of nanocomposite film is ascribed to the radical capturing ability of C_{60} [5-11].

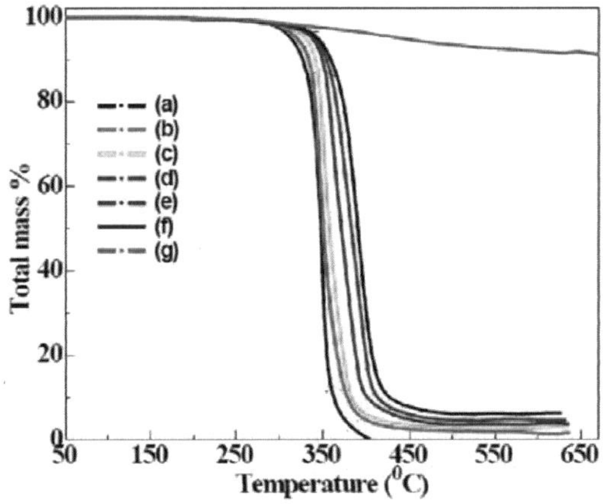

Figure 6. TGA curves for (a) PVP and PVP-C_{60} nanocomposites consisting of (b) 4.63, (c) 9.26, (d) 13.9, (e) 18.52, and (f) 23.15 μM C_{60}. 'g' is for C_{60} without PVP

The macroradicals which were usually obtained during the thermal degradation process are detained by C_{60} so that it restricts the thermal degradation of the polymer. In section 3.1, I report lowest PDI 0.272 for sample (d) consisting of 13.9 μM C_{60}. This shows that in this sample C_{60} molecules are uniformly distributed in PVP matrix and this is reflected as a crest near 13.9 μM C_{60} in Figure 7a. The microscopic image also supports our observation (see Figure 8). In a report, Fernandes et al. [8] have reported an improvement in the thermo-oxidative stability of the functionalized C_{60}/PS composites compared to the unfunctionalized C_{60} composites and they attributed it to the fact that functionalized C_{60} disperses better in a polymer matrix than C_{60}. I also studied the % total residual mass at 550 ^{0}C as a function of C_{60} content (Figure 7b). The char residue % at this temperature is found to be more for the sample containing more carbon materials i.e., C_{60}. The plot shows that two parameters vary linearly with each other and reveals that thermal stability of nanocomposites depends on the content of cross-linkers i.e., the carbon materials.

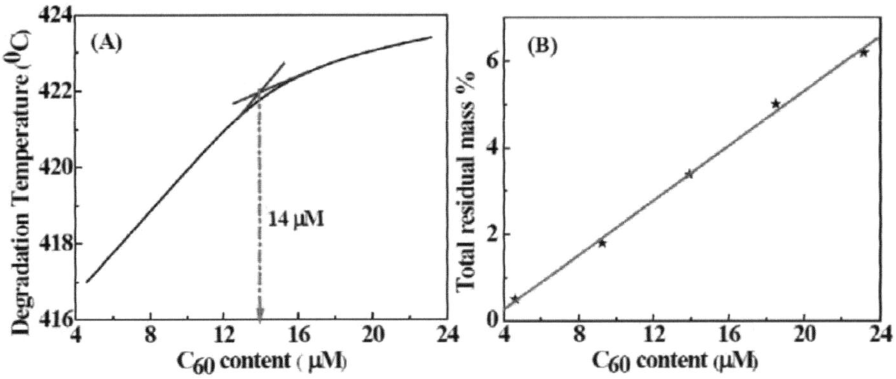

Figure 7. Variation of (A) degradation temperature and (B) total residual mass % against C_{60} content in PVP-C_{60} nanocomposite films.

3.4 Microstructures of C_{60}:PVP nanocomposite films

Figure 8 shows TEM images obtained for C_{60}:PVP nanocomposite films consisting of (a) 4.63, (b) 9.26, (c) 13.9, and (d) 23.15 μM C_{60}. Assemblies of the C_{60} NPs capped by PVP of near spheroids of 10–20 nm diameters are by and large shown in different polygons.

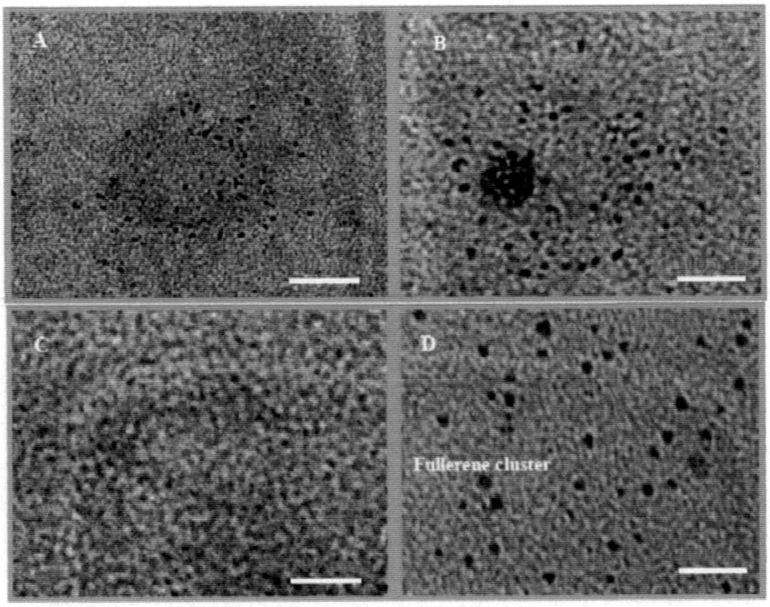

Figure 8. TEM images of PVP-C_{60} nanocomposite films consisting of (a) 4.63, (b) 9.26, (c) 13.9, and (d) (d) 23.15 μM C_{60}.

The C_{60} cluster size increases from 10 nm in sample (a) to nearly 20 nm in sample (d) due to a relative decrease in the encapsulant PVP content. The micrograph for the sample consisting of 13.9 μM C_{60} which shows a uniform distribution of C_{60} NPs in the PVP matrix possess highest degradation temperature (~ 422 ^{0}C) amongst all samples (Figure 8c) owing to possession of suitable network structure.

4.0 Conclusion

From the DLS study I found that an average L_{hd}-value of 9.2 nm for a C_{60}:PVP NF consisting of 4.63 μM C_{60} has grown-up significantly to 17.0 nm when the dopant concentration has increased to 23.15 μM C_{60}. Both L_{hd} and PDI values sketch a non-linear path with the C_{60} doping in the NFs over a wide range of 4–24 μM C_{60} covered in this experiment. In comparison to a negative ξ-value of (-) 5.6 mV with fwhm ~9.1 mV and σ_{sc}-value 0.043 mS/cm in the bare PVP molecules, a relatively broader zeta band of an average ξ-value increased to (-) 12.4 mV, with fwhm ~14.4 mV and an enhanced σ_{sc}-value to 0.065 mS/cm in a C_{60}:PVP NF with 4.63 μM C_{60} reveals accumulation of n-electrons of >N–C=O moieties from the PVP molecules on the C_{60} surface in C_{60}:PVP charge transfer complex. Considerably increase in the thermal stability of nanocomposite films is credited to C_{60} for its excellent radical capturing ability.

Acknowledgements

I highly acknowledge the support of Silicon Institute of Technology, Bhubaneswar, India in carrying out this work.

IOP Conf. Series: Materials Science and Engineering **410** (2018) 012009 doi:10.1088/1757-899X/410/1/012009

References

1. A, Dasari, Z.Z. Yu, Y.W. Mai, 2016, *Polymer Nanocomposites: Towards Multi-Functionality.* Springer.
2. P.M. Visakh, B. N. Olga, 2015, *"Thermal Degradation of Polymer Blends, Composites and Nanocomposites." In Thermal Degradation of Polymer Blends, Composites and Nanocomposites*, Pp. 1-16. Springer International Publishing.
3. M. Prato, 2009, *Fullerene polymers: synthesis, properties and applications*. Eds. N. Martín, F. Giacalone, John Wiley & Sons.
4. C. N. Kramer, ed. 2007, *Fullerene Research Advances*. Nova Publishers.
5. L. Zhao, Z. Guo, S. Ran, Z. Cao, 2014, Fang, *Journal of Thermal Analysis and Calorimetry*, Vol. *115, No.* 2, Pp.1235-1244.
6. H. Gaspar, L. Fernandes, P. Pereira, G. Bernardo, 2015, *Polymer Bulletin*, *Vol.* 7, Pp.1775-1786.
7. R. Katiyar, D.S. Bag, I. Nigam, 2013, *Thermochimica Acta*, 2013, *557*, pp.55-60.
8. L. Fernandes, H. Gaspar, G. Bernardo, *Polymer Testing*, 2014, Vol. *40*, Pp.63-69.
9. K. Kokubo, R. Takahashi, M. Kato, A. Harada, T. Noguchi, T. Oshima, 2016, *Polymer Composites*. Vol. 37, No. 4, Pp.1143-1151.
10. P.A. Song, Y. Zhu, L. Tong, Z. Fang, 2008, *Nanotechnology*, Vol. *19*, No. 22, p. 225707.
11. V.V. Zuev, F. Bertini, G. Audisio, 2005, *Polymer degradation and stability*,Vol. *90*, No.1, Pp.28-33.
12. J.Y. Lee, Y. Liao, R. Nagahata, S. Horiuchi, 2006, *Polymer*, Vol. *47*, No. 23, Pp.7970-7979.
13. T. Cosgrove, 2010, *Colloid science: Principle, methods, and applications*, Wiley.
14. R. J. Hunter, 1998, *Zetapotential in colloid science: Principles and applications*, Academic Press, UK.
15. K. Rahme, L. Chen, R.G. Hobbs, M.A. Morris, C. O'Driscoll, J.D. Holmes, 2013, *Rsc Advances*, Vol. *3*, No.17, Pp.6085-6094.
16. W. Brown, 1993, *Dynamic Light Scattering: The Method and some Applications*, New York: Claredon.
17. B.J. Berne, R. Pecora, 2003, *Dynamic Light Scattering: with Applications to Chemistry, Biology, and Physics*, Dover publications,
18. R.G. Alargova, S. Deguchi, K. Tsujii, 2001, *Journal of the American Chemical Society*, Vol. *123, No.* 43, Pp. 10460-10467.
19. V.R. Chaudhari, S.K. Haram, S.K. Kulshreshtha, J.R., Bellare, P.A. Hassan, 2007, *Colloids and Surfaces A: Physicochemical and Engineering Aspects*, 2007, Vol. *301, No.* 1, Pp. 475-480.
20. G. Mountrichas, S. Pispas, E. Xenogiannopoulou, P. Aloukos, S. Couris, 2007, *The Journal of Physical Chemistry B*, 2007,Vol. *111, No.* 17, Pp. 4315-4319.
21. N.P. Yevlampieva, S.K. Filippov, T.S. Dmitrieva, I.I. Zaitseva, E.Y. Melenevskaya, O.V. Nazarova, E.I. Ryumtsev, 2007, *Polymer Science Series A*, *Vol. 49, No.* 6, Pp. 642-650.
22. Z. Wang, B. Tan, I. Hussain, N. Schaeffer, M.F. Wyatt, M. Brust, A.I. Cooper, 2007, *Langmuir*, Vol. *23, No.* 2, Pp. 885-895.
23. A. Mishra, A., S. Ram, G. Ghosh, 2009, *The Journal of Physical Chemistry C*, Vol. *113, No.* 17, Pp.6976-6982.
24. M. Behera, 2015, *International Nano Letters*, Vol. *5, No.* 3, Pp.161-169.
25. M. Behera, S. Ram, 2015, *Applied Nanoscience*, 2014, Vol. *4, No.*2, Pp. 247-254.
26. M. Behera, S. Ram, 2013, *International Nano Letters*, Vol. *3*, No. 1, p. 1.
27. M. Behera, S. Ram, 2013, *Applied Nanoscience*, Vol. *3, No.* 1, Pp. 83-87.

28. M. Behera, S. Ram, 2015, *Fullerenes, Nanotubes and Carbon Nanostructures*, Vol. *23*, No. 10, Pp. 906-916.
29. M. Behera, 2015, *Research Journal of Nanoscience and Nanotechnology*, Vol. 5, Pp. 60-73.
30. M. Behera, S. Ram, 2016, *Plasmonics*, Vol. 14, No. 4, pp. 1057-1065.

IOP Conf. Series: Materials Science and Engineering **410** (2018) 012010 doi:10.1088/1757-899X/410/1/012010

Evolution of structural and magnetic properties of Ti doped Sr_2IrO_4: a novel magnetic insulator

Imtiaz Noor Bhatti[1] and A. K. Pramanik

School of Physical Sciences, Jawaharlal Nehru University, New Delhi-110067, India

ABSTRACT

5*d* transition metal oxides (TMOs) recently caught attention of researchers because of potential to show exotic phenomenon. Crystal field effect (CFE), spin-orbital interaction (SOI) and onsite Coulomb interaction (U) in 5d oxides are on comparable scale and their interplay among themselves setting new balance of energies and drive exotic physics in these materials. Among several 5d oxides iridates are extensively studied, where Sr_2IrO_4 is prototype of novel physics and stabilize in J_{eff} = ½ ground state. The Sr_2IrO_4 is first member of Rudellson-Popper series $Sr_{n+1}Ir_nO_{3n+1}$ with n =1 and crystallize in reduced tetragonal structure and adopt *$I4_1/acd$* space group. The physical properties of Sr_2IrO_4 are influenced by this reduced symmetry due to rotation of IrO_6 octahedral. The magnetic ground state in Sr_2IrO_4 is canted type antiferromagnetic (AFM), it's this canted spin structure which gives ferromagnetic character to this material with ordering temperature around 225 K to this material. The spin canting is rendered by Dzyaloshinskii-Moriyam anti-symmetric interaction driven by SOI and rotated IrO_6 octahedra. We endeavour to perturb the interactions viz. SOI and U in Sr_2IrO_4 by substituting Ti^{4+} ($3d^0$ S = 0) for Ir^{4+} ($5d^5$ J_{eff} = 1/2). We have studied the effect of Ti doping on structural and magnetic properties. We observe evolution in unit cell lattice parameters, in the meantime Ir-O-Ir bond angle also increases which reduces octahedral rotation (θ_{Oct}). Further, our magnetization study shows decrease in transition temperature (T_C), increase in effective magnetic moment and doping influence the coercively and remnant magnetization. Thus we observe Ti doping prominently effect the structural and magnetic properties in Sr_2IrO_4. The results presented here provide insight into the effect of Ti doping on structure and magnetic phase in strong spin-orbit coupled Sr_2IrO_4.

Keywords : Doping, Insulator, magnetization, spin-orbit coupling

Corroseponding Author: inbhatti07@gmail.com

1. INTRODUCTION

Recently 5d-based iridates and osmates have received much attention of researchers due to exotic physics arising from competing interactions viz. crystal field effect (CFE), spin-orbital interaction (SOI) and onsite Coulomb interaction (U).[1-5] Among them the most extensively studied iridate is the layered perovskite Sr_2IrO_4, believed to stabilize in J_{eff} = ½ ground state and it is a spin orbital coupling driven Mott insulator. Sr_2IrO_4 having similarities with high T_C cupric superconductor La_2CuO_4 draw interest, as it has also been theoretically predicted to be superconductor electrons/hole doping. In *5d* iridates active interactions such as: (CEF), (U) and (SOI), compete among themself setting new level of energies and indorse novel magnetic phases in these materials. It is recently found that the ground state is driven by strong SOI of Ir ion in an octahedral environment of oxygen atoms which splits the t_{2g} band into lower $J_{eff} = 3/2$ quartet and upper $J_{eff} = 1/2$ doublet. The $J_{eff} = 1/2$ band is so narrow that even reduced U opens a Mott gap by splitting it into upper and lower Hubbard bands, thus driving it to unconventional Mott insulator.[1-2] The Sr_2IrO_4 crystallize in reduced tetragonal structure with space group $I4_1/acd$. The reduced symmetry is due to rotation of IrO_6 octahedral around c-axis and play important role in deciding the physical properties of Sr_2IrO_4. [1-7]

The magnetic ground state in Sr_2IrO_4 is believed to be canted type antiferromagnetic (AFM) which gives ferromagnetic component with magnetic ordering around 225 K to this material. [5-11] The magnetic ground state is quite delicate and can be perturb and destabilized by tuning the any of the interaction band filling, SOC and U. To achieve this motivation chemical doping is often employed. Here we have doped Ti^{4+} as substitution for Ir^{4+} which introduce holes and reduces the SOC while increase U.

In this paper we investigate the evolution of structural and magnetic properties of $Sr_2Ir_{1-x}Ti_xO_4$ (x = 0.0, 0.05 and 0.1). We observe evolution of structure parameters with Ti doping bot no structure phase transition is observed. The magnetization is studied in detail; all samples are ferromagnetic in nature. However, the transition temperature decreases with doping, where as effective magnetic moment increases. The isothermal magnetization shows decrease in coercively and remnant magnetization.

2. EXPERIMENTAL DETAIL

The polycrystalline samples of $Sr_2Ir_{1-x}Ti_xO_4$ with composition x = 0.0, 0.05 and 0.1 are prepared using slandered solid state reaction technique. The starting material IrO_2 (Sigma Aldrich 99.99 %), TiO_2 (Sigma Aldrich 99.99 %), and $SrCO_3$ (Sigma Aldrich 99.95 %), are taken in stoichiometric composition and grinded well. The fine powder is then calcinated at 900 ^{0}C for one day. The calcinated powder is sintered twice at 1000 ^{0}C and 1100 ^{0}C with intermediate grindings, the detail procedure is described elsewhere.[9,10,11] Samples in powder form are characterized by x-ray diffraction (XRD). The structural analysis is done on XRD data using FullProof software. DC magnetization (M) data have been collected using a vibrating sample magnetometer (PPMS, Quantum Design).

IOP Conf. Series: Materials Science and Engineering **410** (2018) 012010 doi:10.1088/1757-899X/410/1/012010

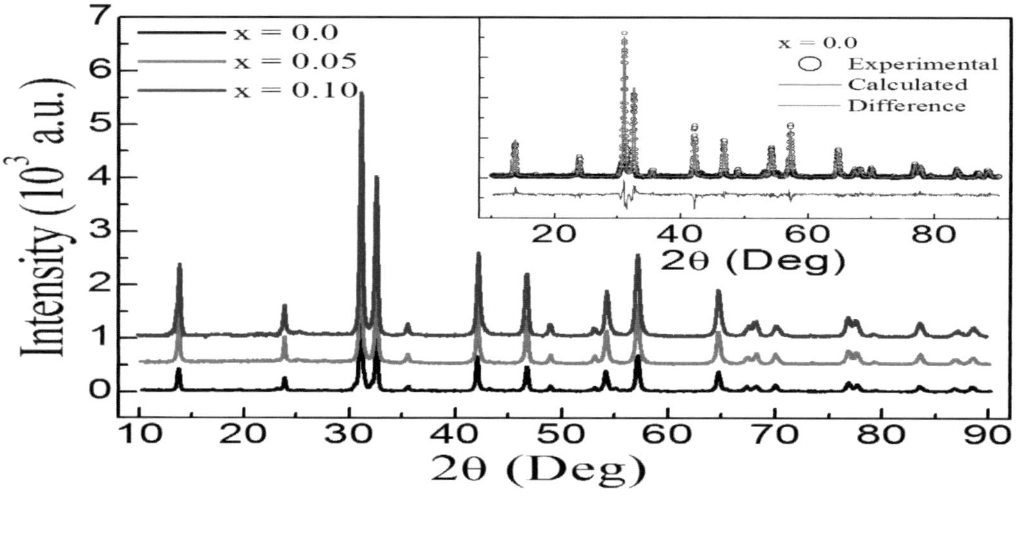

Figure 1. X-ray diffraction data collected at room temperature for $Sr_2Ir_{1-x}Ti_xO_4$ sample series. Inset shows XRD pattern of Sr_2IrO_4 along with the Reitveld refinement.

3. RESULTS AND DISCUSSION

3.1. Structural study

Fig. 1 shows the XRD pattern of $Sr_2Ir_{1-x}Ti_xO_4$ sample series, we observe that there is no extra peak present in doped samples which shows samples are in single phase. Inset Fig. 1 shows Rietveld refinement of parent sample with x =0.0. The structural analysis on all samples shows the samples crystalizes in single phase and adopts tetragonal structure with $I4_1/acd$ space group. The refined lattice parameters are $a = b = 5.4980(2)$ Å and $c = 25.779(1)$ Å for parent sample, with Ti doping lattice parameter evolve slightly. We also find a rotation of IrO_6 octahedra decreases from ~ 11.3° for x = 0.0 to 8° for x = 0.1 around c axis.

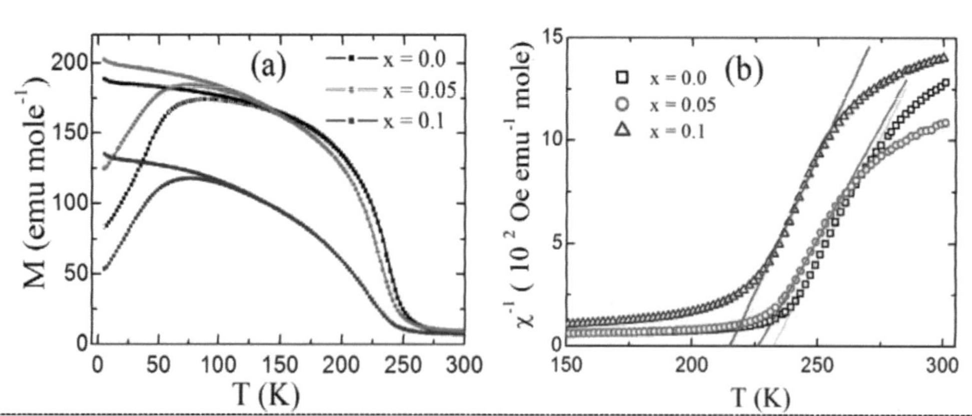

Figure 2. (a) ZFC and FC magnetization data measured in 10 kOe are plotted as a function of temperature for $Sr_2Ir_{1-x}Ti_xO_4$. (b) Inverse susceptibility (χ^{-1}) shows Curie-Weiss behaviour.

3.2. Magnetization Study

Fig. 2(a) shows the temperature dependant magnetization M(T) data collected in ZFC and FCW protocol in the temperature range of 5 K – 300 K in an applied field of 10 kOe. The M(T) data show a phase transition from paramagnetic (PM) to weak ferromagnetic (FM) state with $T_C \sim 225$ K for the parent compound. However, with Ti doping moment and T_C decreases. Fig. 2(b) shows the invers susceptibility as a function of temperature. In PM state the susceptibility is fitted with Curie-Weiss law expressed as:

$$\chi = \frac{C}{T - \theta_P}$$

Where χ is magnetic susceptibility, C is the Curie constant and θ_P is the Curie temperature.
We have calculated effective paramagnetic moment using fitting parameters for all samples $\mu_{eff} = 0.56$ μ_B/f.u. for parent compound. With Ti doping μ_{eff} increases which reflect the weakening of exchange interaction, making spins free to align in field direction easily.

To further understand the magnetic properties we measured isothermal magnetization M(H) at T = 5 K shown in Fig. 3(a). The hysteresis in M(H) is suggestive of FM ordering in this material. At 5 K, the magnetic moment $\mu_S = 0.05$ μB/f.u at H = 70 kOe is observed, which decreases with Ti doping. It is quite evident from the Fig. 3(a) that corecivity and reminance decreases with doping. Fig. 3(b) shows the Arrott plot (M^2 vs H/M) ,the extrapolation of Arrott plot at high field intercept the positive M^2 axis which realize the presence of spontaneous moment in these samples.

The suppression of transition temperature TC and moment can be explain by considering the effect of Ti substitution since Ti^{4+} is S = 0 is a non-magnetic and hence break the Ir-O-Ir magnetic channel thus weaken the exchange interactions and suppress the T_C and decreases magnetic moment.

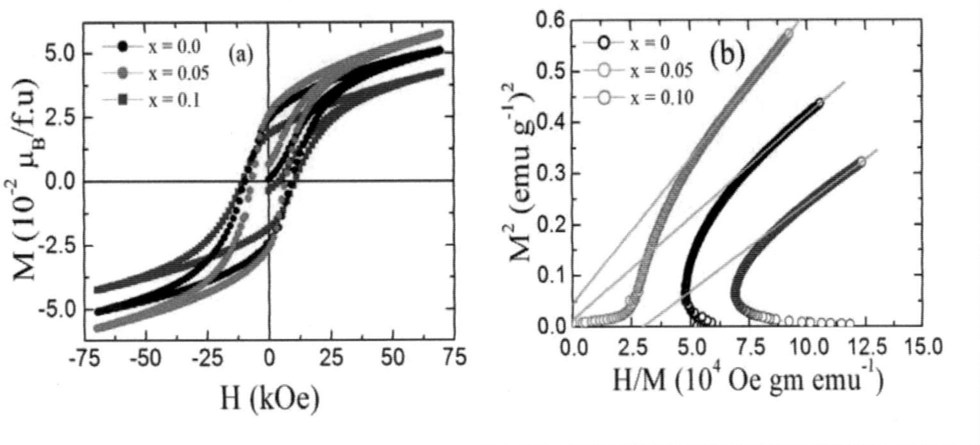

Figure 3. (a) Isothermal magnetization is shown for $Sr_2Ir_{1-x}Ti_xO_4$ at 5 K. (b) Arrott Plot (M_2 vs H/M) solid lines due to extrapolation at high field.

4. CONCLUSIONS

We have prepared single phase polycrystalline samples of $Sr_2Ir_{1-x}Ti_xO_4$. Rietveld analysis of XRD data shows samples adopt tetragonal structure with crystal structure $I4_1/acd$ space group. The detail temperature dependent magnetization M(T) study shows the moment decreases with Ti doping and suppression in T_C. Ti doping breaks the Ir-O-Ir chain thus weakens the exchange interaction. M(H) data shows that all samples are ferromagnetic in nature. However, coercivity and remnant magnetization decreases with doping.

ACKNOWLEDGMENTS

We acknowledge UGC-DAE CSR, Indore for measurement facilities. One of the authors (INB) acknowledges the CSIR-India for fellowship.

REFERENCES

[1] B.J. Kim, Hosub Jin, S.J. Moon, J.-Y. Kim, B.-G. Park, C.S. Leem, Jaejun Yu, T.W. Noh, C. Kim, S.-J. Oh, J.-H. Park, V. Durairaj, G. Cao, and E. Rotenberg, 2008, Physical Review Letter, Vol. 101, pp. 076402.
[2] S.J. Moon, H. Jin, K.W. Kim, W.S. Choi, Y.S. Lee, J. Yu, G. Cao, A. Sumi, H. Funakubo, C. Bernhard, and T.W. Noh, 2008, Physical Review Letter, Vol. 101, pp. 226401.
[3]M.K. Crawford, M.A. Subramanian, R.L. Harlow, J.A. Fernandez-Baca, Z.R. Wang, and D.C. Johnston, 1994, Physical Review B, Vol. 49, 9198.
[4] M. Ge, T.F. Qi, O.B. Korneta, D.E. De Long, P. Schlottmann, W.P. Crummett, and G. Cao, 2011, Physical Review B, Vol. 84, pp. 100402.
[5] J.S. Lee, Y. Krockenberger, K.S. Takahashi, M. Kawasaki, and Y. Tokura, 2012, Physical Review B, Vol. 85, pp. 035101.
[6] T.F. Qi, O.B. Korneta, L. Li, K. Butrouna, V.S. Cao, X. Wan, P. Schlottmann, R.K. Kaul, and G. Cao, 2012, Physical Review B, Vol. 86, pp. 125105.
[7]T. Shimura, Y. Inaguma, T. Nakamura, M. Itoh, and Y. Morii, 1995, Physical Review B, Vol. 52, pp. 9143.
[8] S. Calder, G.-X. Cao, M.D. Lumsden, J.W. Kim, Z. Gai, B.C. Sales, D. Mandrus, and A.D. Christianson, 2012, Physical Review B, Vol. 86, pp. 220403.
[9]I.N. Bhatti, R. Rawat, A. Banerjee, and A.K. Pramanik, 2014, Journal of Physics: Condense Matter, Vol.27. No. 1, pp. 016005.
[10]I.N. Bhatti and A K Pramanik, 2017, Journal of Magnetism and Magnetic Material, Vol. 422, pp. 141.
[11] I.N. Bhatti and A.K. Pramanik, 2017, American Institute of Physics Conference Proceedings, Vol. 1731, pp. 110046.
[12]I.N. Bhatti, R.S. Dhaka, and A.K. Pramanik, 2017, Physical Review B, Vol. 96. No. 14, pp. 144433.

Graft copolymerization of Soy Protein Isolate with Polylactide via Ring Opening Polymerization

Sasmita Pattanaik[1], Alekha Kumar Sutar[1] and Tungabidya Maharana[2*]
[1]Catalysis Research Lab, Department of Chemistry, Ravenshaw University, Cuttack-3, Odisha, India
[2]Department of Chemistry, National Institute of Technology, Raipur, India
Corresponding author: Dr. T Maharana; Tel-+91-771-2254200; Fax: +91-771-2254600; E-mail: mtungabidya@gmail.com

Abstract

The present work is devoted to the graft copolymerization of polylactide onto soy protein isolate backbone by ring-opening polymerization in the presence of $Sn(Oct)_2$ catalyst. The effect of various input parameters on the output parameters has been carried out to investigate the required levels to get an optimum product. Maximum graft yield of 81.4% polylactide-g-soy protein isolate could be obtained at 90°C within 12 h and while using SPI amount 1.5g, 1g lactide and preswelling time of 14 h. Characterizations of the grafted polymers were carried out by various methods like Fourier transform infrared (FTIR) spectroscopy, Nuclear Magnetic Resonance (NMR) spectroscopy, thermal (TGA/DTA/DSC), X-ray Diffraction (XRD) and Field emission Scanning electron microscopy (FESEM). The swelling behavior of the SPI and grafted copolymers were studied in binary solvent (water: ethanol) system and the % swelling and % solubility has been determined.

Keywords: Ring opening polymerization, polylactide, soy protein isolate, biodegradable polymer, graft copolymerization.

Introduction

Researchers around the world have made significant technological advance by synthesizing biodegradable polymers like polylactic acid (PLA), polycaprolactone (PCL), poly(hydroxybutyrate) (PHB) and poly(hydroxybutyrate - valerate) (PHBV).[1,2] Its demand has increased as a replacement to synthetic thermoplastic for various applications viz. controlled release of medicaments into the human and animal bodies, manufacture of bioabsorbable prostheses or controlled release of insecticides in the agricultural field. However, these polymers are usually more expensive than petroleum - based polymers. Thus, to solve this problem, development of low cost biodegradable polymers has attracted a number of research scientists for last three decades.[3-4]

Therefore, in the present investigation soy protein isolate (SPI) has been chosen which is a low cost and profusely obtainable agronomic material, which can become a good material for manufacturing biodegradable plastics and food and/or non-food packaging films. However, soy protein could not be processed without a sufficient amount of plasticizers like common thermoplastic materials because of strong intra- and intermolecular interactions, which lead to its low mechanical properties.[5] Further, high polarity and hydrophilicity of soy protein molecules restricts the applications of soy protein in plastic materials. Thus, in order to overcome the above mentioned issues, soy protein has been blended with biodegradable polyesters and grafted by some polymers such as polyethylene, polylactic acid, polydioxane etc. However, difference in polarities makes most of the polyesters, amide and esteramides immiscible with protein.[3-4] Subsequently, decrease in the physical behaviour is observed due to thermodynamic incompatibility and poor interfacial adhesion.

Based on literature review, a very few information about graft copolymerization of soy protein with cyclic esters has been found.[5] According to Liu research group the aliphatic polyesters such as poly(*p*-dioxanone) PDO, polycaprolactone PCL, polylactide PLA, etc, has been used in medical application and chemical industries for their biodegradability, biocompatibility and good mechanical behaviour.[6-8] Soy protein becomes water resistance with thermoplasticity by grafting/blending with aliphatic polyesters due to their hydrophobicity and excellent processability nature.

Therefore, in nutshell, the present work is devoted to the graft copolymerization of polylactide onto soy protein isolate by ring-opening polymerization in the presence of $Sn(Oct)_2$ catalyst. The effect of various input parameters on the output parameters has been investigated so as to control these in required levels to get an optimum product. The structure of graft copolymers were analysed by Fourier transform infrared (FTIR) spectroscopy, ^1H NMR, differential scanning calorimetry (DSC), thermogravimetry (TG), and wide-angle X-ray diffraction (WAXD).

IOP Conf. Series: Materials Science and Engineering **410** (2018) 012011 doi:10.1088/1757-899X/410/1/012011

Experimental

Materials

Soy protein isolate (SPI) with a protein content of 95% (dry basis) was purchased from Adeesh Agrofoods, Haryana, India. 3,6-dimethyl-1,4-dioxane-2,5-dione (L-Lactide) was purchased from Sigma Aldrich (Netherland) and was used as received and were used as monomer. Stannous octoate ($Sn(Oct)_2$) was purchased from Sigma Aldrich and was used as received. Toluene was received from Himedia Laboratory Pvt. Ltd, India. It was dried by refluxing over a Na/benzophenone complex, and was distilled under nitrogen just before use. Cyclohexane was purchased from Nice Chemicals Pvt Ltd, Cochin and was used without further purification.

Synthesis of polylactide-g-soy protein isolate

The graft copolymerization of polylactide onto soy protein isolate was carried out under purified nitrogen atmosphere. A 50 ml three-necked round-bottomed flask was taken as the reactor, requisite amount of soy protein isolate and Millipore water was poured into the reactor and was placed inside silicone oil bath maintained at required temperature. It was fitted with gas inlet & outlet tubes at 2 extreme necks and the condenser was fixed at the middle neck and purified nitrogen gas was purged through the reactor. Preswelling of the SPI molecules was conducted with stirring of the mixture at 80°C for particular time period and further refluxed under nitrogen atmosphere by using Millipore water. After removal of the water in the mixture by azeotropic reflux with cyclohexane for 4 h, monomer L-lactide and Sn $(Oct)_2$ (0.3 wt % of total amount of the reactants) was added, and graft copolymerization reaction proceeded at various reaction temperature for varying time period. Finally, after the desired time period, the products were purified in a Soxhlet extractor with acetone for 72 h to remove the unreacted monomers and the oligomers of Polylactide and then dried under vacuum to a constant weight.

The graft yield (%) was calculated as the percentage increase in weight of dry precipitate over the original weight of soy protein isolate.[9] The percentage of grafting was calculated using Fanta's formula as given in Eq. 1.

$$\text{Percentage of graft yield} = \frac{\text{Weight of polymer in graft}}{\text{Weight of ungrafted polymer}} \times 100 \quad ....(1)$$

$$= \frac{\text{Wt of grafted polymer} - \text{Wt of ungrafted polymer}}{\text{Wt of ungrafted polymer}} \times 100$$

$$= \frac{\text{Wt of grafted copolymer} - \text{Wt of ungrafted SPI}}{\text{Wt of ungrafted SPI}} \times 100$$

Characterization Methods

Soy protein isolate and polylactide-g-SPI sgrafted (graft yield-81.4%) samples were characterized by using different analytical instruments such as FTIR (Thermo Nicolet Avatar 370) spectroscopy, ^{1}H NMR spectroscopy (Bruker Avance 500), X-Ray Diffractometer (XRD) Bruker AXS D8, Thermal analyzer (TGA/DTG) (TGA Q-50 TA Instruments), Differential Scanning Calorimetry (DSC) DSC Q-20 TA Instruments and Field Emission Scanning Electron Microscopy (FESEM) QUANTA 200FEG.

Swelling Study of Polylactide-g-Soy Protein Isolate

Swelling behaviour of neat soy protein isolate and grafted copolymers of polylactide-g-soy protein isolate was studied in a polar solvent to determine the swelling percentage and percent solubility.

Procedure for Swelling Study

1 g sample each of neat SPI and grafted copolymers of polylactide-g-soy protein isolate (graft yield-81.4%) taken separately were suspended in 10 ml of the binary polar solvent system of mixture of ethanol and water in different proportion and the reaction vessel was kept at 40°C. The samples were filtered after 24h. The adhered water was dried by pressing the sample gently between the folds of filter paper and weighed immediately. The percent swelling was determined from the increase in the weight of the original sample as follows:

$$\% \text{ Swelling} = \frac{W_s - W_0}{W_0} \times 100$$

Where W_s is the weight of swollen polymer and W_0 is the weight of original polymer. The swollen sample was dried in an oven till constant weight is obtained and percent solubility was calculated from the following equation:

$$\% \text{ Solubility} = \frac{W_d - W_0}{W_0} \times 100$$

Where W_o is the weight of original polymer and W_d is the weight of dried sample after swelling. Dried samples were again placed for swelling study in the solvent for another 24h at the same temperature. Percent swelling and percent solubility was calculated as follows:

$$\% \text{ Swelling} = \frac{W_{d'} - W_d}{W_d} \times 100$$

Where W_d is the weight of dried polymer after 24h and $W_{d'}$ is the weight of swollen polymer sample after 48h.

$$\% \text{ Solubility} = \frac{W_{d''} - W_{d'}}{W_{d'}} \times 100$$

Where $W_{d'}$ is the weight of dried polymer after 24 h and $W_{d''}$ is the weight of dried sample after swelling for 48 h.

Results and Discussion

Copolymerization of polylactide-g-SPI

The total numbers of operating parameters that influence the synthesis of polylactide-g-soy protein isolate grafted copolymers are about six. The effects of various parameters such as reaction time, reaction temperature, preswelling time, concentration of soy protein isolate, lactide, and catalyst stannous octoate on grafting were studied to optimize the reaction conditions for maximizing graft copolymerization. The effects of various parameters were studied by varying one parameter and keeping all other parameters constant. The most significant output parameter for synthesis of polylactide-g-soy protein isolate grafted copolymers is graft yield (wt. %) and it regulate physical, mechanical & biodegradable properties of the polymers, respectively.

Effect of reaction time on graft yield

Fig. 1 shows the influence of the reaction time on the graft copolymerization. It has been observed from Figure 1 that with the increase of the reaction time 7-12 h, the percentage of grafting increases and with further increase in reaction time period, graft yield (%) decreases. Thus, the optimum time period for the graft copolymerization process is around 12 h.

The initial increase in graft yield (%) with increasing time period is attributed to the fact that with increasing reaction time, the concentration of initiating species increases and with further increasing in time period, destruction of initiating species occurs and thereby leading to decreasing graft yield (%) with time period.[10] The process of depolymerization occurs with longer time period and thereby decreasing the graft yield (%) after the optimum time period of 12 h. Beyond the optimum time period, the active sites on polylactide and SPI might be killed leading to decrease in percent graft yield.[4]

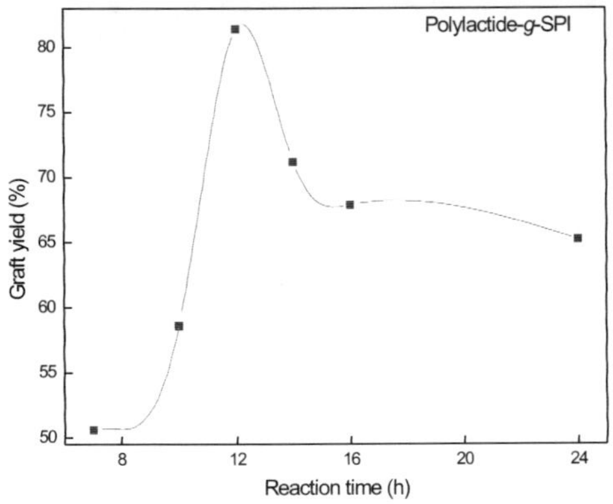

Fig. 1 Variation of graft yield (%) of Polylactide-g-SPI with Reaction Time

Effect of reaction temperature on graft yield

The effect of reaction temperature on graft yield (%) was studied within the temperature range 80-110°C and is depicted in Fig. 2. From the Fig. 2, it is evident that with the increase in temperature from 80°C to 90°C, the graft yield (%) has increased but on further increasing the temperature the graft yield (%) was found to decrease. The percentage of graft yield was maximum at 90°C.

This can be explained by assuming that on increasing the temperature from 80-90°C, solubility of PLA, swellability of SPI, attack of PLA units on the SPI backbone in the solution favors grafting and thereby increases the graft yield (%). But above 90°C, the interaction of SPI macro radicals with Sn(II) ions destroys the activity of the initiating species and thus leading to decreased graft yield (%) at higher temperatures. [4,5,11]

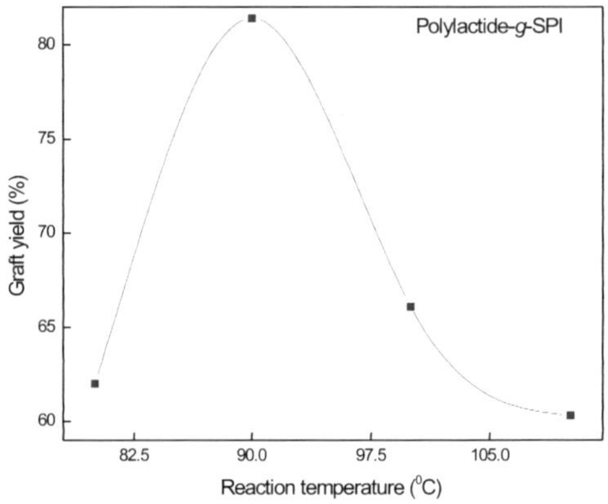

Fig. 2 Variation of graft yield (%) of Polylactide-g-SPI with Reaction Temperature

Effect of amount of SPI on graft yield

The effect of amount of soy protein isolate (SPI) on graft yield (%) is given in Fig. 3. It was observed that on increasing the amount of SPI from 0.5 to 1.5 g the percentage of grafting increases and with further increase in amount of SPI, graft yield (%) is found to decrease.

The increasing trend may be due to increasing concentration of SPI macro radicals at higher concentration and the rate of their combination with lactide units has become faster. However, at further higher concentration of SPI, the graft yield (%) decreases which may be attributed to the fact that the concentration of SPI macro radicals increases further with increase in concentration of SPI and the rate of their combination and disproportion are faster than the rate of their combination with polylactide (PLA) molecules. Further, the rate of diffusion of PLA onto the surface of SPI decreases with further increase in concentration of SPI which hinders the grafting of PLA. Alternatively, when the concentration of the SPI was above a certain value, there were more free radicals attacking the lactide monomeric or polymeric units than the SPI and thus, more PLA homopolymer is formed. [4,5,10]

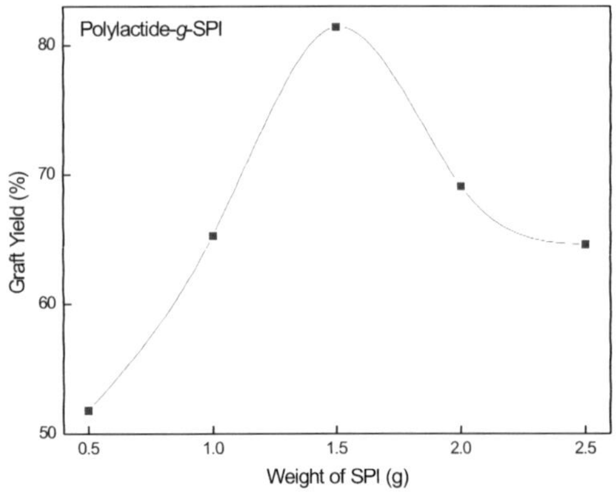

Fig. 3 Variation of graft yield (%) of Polylactide-g-SPI with Weight of SPI

Effect of concentration of L-Lactide on graft yield

The variation plot of concentration of L-Lactide (LA) with graft yield is shown in Figure 4. It was observed from the Figure that on increasing the amount of L-Lactide (LA) from 0.5 to 1.0 g the percentage of grafting increases and with further increase in amount of LA graft yield (%) decreases.

The initial increase in the graft yield percentage was obviously due to the greater availability of monomer in the proximity of the SPI macroradicals. At higher concentration of L-Lactide, the decrease in graft yield may be due to the wastage of LA molecules in the formation of a large amount of homopolymer. $Sn(Oct)_2$ can react with a trace amount of water to form the active derivatives, which promotes the lactide polymerization directly as PLA homopolymer.[4,5] Moreover, the large molecular weight PLA homopolymer increased the viscosity of the reaction medium. The diffusion coefficient of the monomer molecules to the SPI macroradicals was reduced because of the increase of the viscosity of the reaction system, resulting in the gradual decrease in the percentage of grafting, as well as the percentage of efficiency.[12]

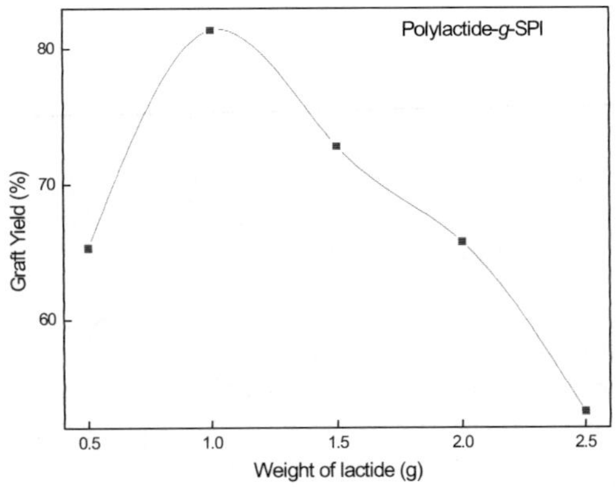

Fig. 4 Variation of graft yield (%) of Polylactide-g-SPI with Weight of L-Lactide

Effect of amount of catalyst on graft yield

The effect of amount of catalyst (wt %), stannous octoate (Sn $(Oct)_2$), on graft yield (%) is shown in Figure 5. From the figure, it was observed that with the increase in amount of catalyst from 0.2 wt % to 0.3 wt %, the graft yield (%) increases and with further increasing to 0.4 wt %, the graft yield (wt%) was found to decrease. The graft yield (wt %) was highest when the amount of catalyst was maintained at 0.3 wt % during the reaction.

With the increase in amount of catalyst from 0.2 to 0.3 wt %, a large number of stannous dihydroxide initiator is formed which helps in opening of the lactide ring, which further interacts with the SPI macroradiacals and initiate grafting and thereby increasing the graft yield.[13] However, with further increase in amount of catalysts to 0.4 wt %, abundant amount of lactide will undergo ring opening and thus will lead to more amount of homopolymer formation. Some of these active species of lactide monomeric units will also interact with SPI macroradicals and will lead to graft coplymerization. Homopolymer macroradicals formed might terminate the growing grafted chain; the active sites formed on the main chain of SPI might be oxidized and thus terminate the reaction site. The process of graft copolymerization and homopolymerization in such a system is a matter of competition, which depends on the direct attack of active species on SPI or on lactide monomer or polylactide units.[4,13] When the concentration of the catalyst was above a certain value, they are mainly favoring the homopolymerization reaction instead of graft copolymerization and thus the graft yield (wt %) is found to decrease with the increase in amount of catalyst.[10,14]

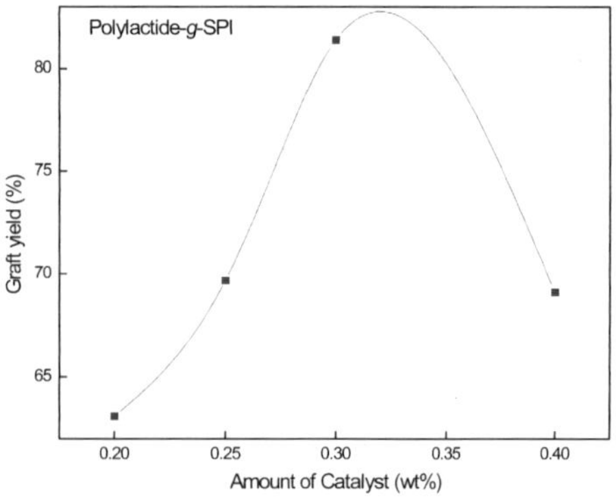

Fig. 5 Variation of graft yield (%) of Polylactide-g-SPI with Concentration of Catalyst

Effect of preswelling time on graft yield

Figure 6 shows the variation of graft yield (%) with peswelling time. It has been observed from the figure and Table mentioned above that with the increase of the preswelling time from 8 to 14 hour, the graft yield (%) increases and then with further increase in the preswelling time, graft yield (%) was found to decrease. It is clear to see that reaction could proceed effectively after the preswelling of protein.

The native globular SPI is an oblate cylindrical structure. Heating the folded SPI causes it to unfold. With the increase of preswelling time, the unfolding structure of SPI increases. These unfolded structures of SPI have more free volumes, chain mobility and exposed active groups, such as -OH, -NH, and so forth, and hence can easily react with PCL. It may further be attributed to the fact that the inter- and intramolecular hydrogen bonding interactions of soy protein decreased after the preswelling and as a result, the soy protein chains became unfolded and therefore promotes the grafting reaction.[4,5]

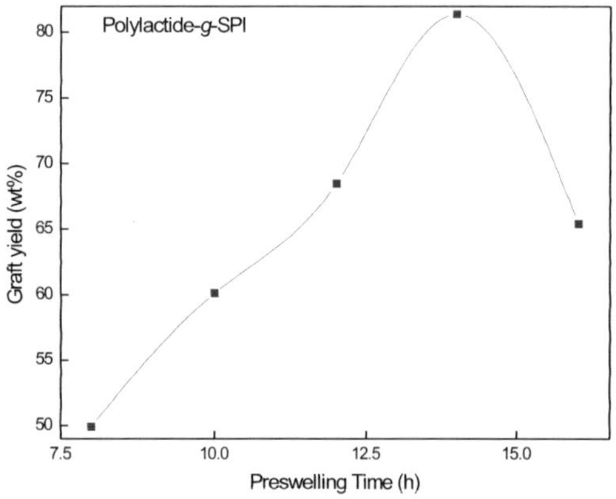

Fig. 6 Variation of graft yield (%) of Polylactide-g-SPI with Preswelling Time

However, if the preswelling time is too long enough, more SPI main chains assembled or self-assembled. With the unfolding of SPI, -OH groups and -NH groups of globulins such as 2*s*, 7*s*, 11*s*, and 15*s* formed strong intermolecular hydrogen bonding during the preswelling process. Therefore, these globulins will assemble or self-assemble. Thus, less active radicals were created, which led to the lower percentage of grafting and efficiency.

Characterization Studies of the polylactide-g-SPI grafted copolymer

In this section, the characterizations of pure SPI and polylactide-g-soy protein isolate grafted copolymer by different techniques such as FTIR, NMR, thermal study, XRD and FESEM are discussed.

FTIR Analysis

For the pure soy protein, the strong band at 3463 cm^{-1} was attributed to the mixed absorption of N-H stretching and O-H stretching. The peak at 1651 cm^{-1} was the contribution of the carbonyl group stretching vibration (amide I band).The band at 1531 cm^{-1} (amide II band) indicated that the N-H had the trans configuration relative to the carbonyl group. In fact, N-H and C=O are trans configurated in all peptide units except proline because less energy is needed in the trans configuration.[5] In addition, the absorption bands related to C-H deformation at 1458 cm^{-1}, C-N stretching and N-H bending vibrations at 1237 cm^{-1}, and out-of-the plane C-H bending at 1064 cm^{-1} were also observed.

The FTIR spectrum of L-lactide depicts a new strong peak at 1757 cm^{-1} for C=O was observed, which was assigned to the absorption of carbonyl in lactide (LA). The peak at 2915 cm^{-1} and 2950 cm^{-1} which belonged to symmetric and asymmetric C-H stretching respectively. Further, the peak at 1267 cm^{-1} is observed for asymmetric stretching vibrations of C-O-C in lactide ring. In FTIR spectrum of Polylactide-g-SPI, characteristic absorption bands at 3305 cm^{-1} for N-H stretching vibration, 2927 cm^{-1} for C-H symmetric stretching vibration due to presence of CH_3 group. The peak at 1741cm^{-1} confirms the presence of C=O group, and peak at 1384 cm^{-1} is for symmetric (C-H) bending vibration.

Characteristic FTIR bands of L-lactide, SPI and Polylactide-g-SPI are depicted in Table 1. The presence of peaks both due to functional groups of SPI and lactide/polylactide in the FTIR spectrum of the grafted product Polylactide-g-SPI indicated that Polylactide chains are present in the copolymer.

Nuclear Magnetic Resonance (NMR) Spectroscopy

Chemical structures of pure SPI and Polylactide-g-SPI were also determined by ^1H NMR spectra in deuterated Chloroform ($CDCl_3$). The NMR spectrum of SPI and Polylactide-g-SPI are shown in Fig. 7 (a,b). Soy protein isolate has a complex composition that results in a complex solution ^1H NMR spectrum (Fig. 7 (a)). SPI contains many kinds of amino acid, mainly glutamic acid, aspartic acid, arginine, and lysine. The amide signals from peptide groups of SPI are displayed at around 7.5 ppm. The aromatic signals of several amino acids are observed at 7.0 ppm, the olefinic signal at 5.0-5.6 ppm and an aliphatic signal at 0.5-5.0 ppm.[5] In the NMR spectrum of SPI, peaks are observed at around 5.3, 4.3, 4.1, 3.5, 2.3, 1.5, 1.3, 0.8 ppm.

Table 1 Characteristic FTIR bands of L- Lactide, SPI, Polylactide-g-SPI

Wave Number (cm^{-1})	Assignment	Ref.
Aliphatic Hydrocarbon		
~3002	-CH asymmetric stretching of CH_3 group in lactide	15
~2927, ~2015	-CH symmetric stretching vibration of CH_3 group in lactide	16
~1390, ~1384	-CH symmetric bending from CH_3group Lactide and PLA-g-SPI	5
~1106, ~1064	C-H bending in SPI	5
Aliphatic Ester		
~1742	C=O stretching vibration for ester group	17
~3511	O-H stretching in PLA	17
~1263	C-O-C symmetric stretching vibration in Lactone ring	5
Aliphatic Amide		
~3463	N-H stretching of amide in SPI	5
~3305	N-H stretching in PLA-g-SPI	5
~1651, ~1531	C=O stretching of amide group in SPI	5

PLA produces mainly three peaks corresponding to the presence of three different kinds of hydrogen atoms in the PLA solution. The three peaks are observed around chemical shift, δ (ppm, $CDCl_3$) 7.25 (s-singlet),

5.15-4.95 (q-quartet, 1H) and 1.60-1.45 (d-doublet, 3H), in the ^{1}H NMR spectrum of PLA. The peaks around 5 ppm and 1.5 ppm are due to the methine (–CH) and methyl (–CH_3) groups of PLA, respectively.[18] The peaks for hydrogen of methine or methyl group of lactide are also observed near the corresponding peak for PLA, but its intensity is very low to be perfectly noticeable.

In the NMR spectrum of Polylactide-g-SPI, the new peaks observed at 5.2 ppm and the complexity of the NMR spectrum near 0.8-2.3 ppm for the amines of SPI and peaks due to polylactide are present. The NMR spectrum of Polylactide-g-SPI (Figure 7) is found to be complex and the peaks present in the grafted are samples are both due the functional groups present in both SPI and polylactide. The new peak at 0.8-1.0 ppm could be attributed to the aliphatic NH-CO group PLA-g-SPI, indicating that the graft mechanism as shown in Scheme 1 is reasonable.[5] The peak at 7.282 ppm (Fig 7(b)) is due to the solvent $CDCl_3$. The amino acids containing benzene ring in SPI were exposed and gave rise to NMR signals after the graft copolymerization. The above information confirms that PLA-g-SPI was synthesized successfully.

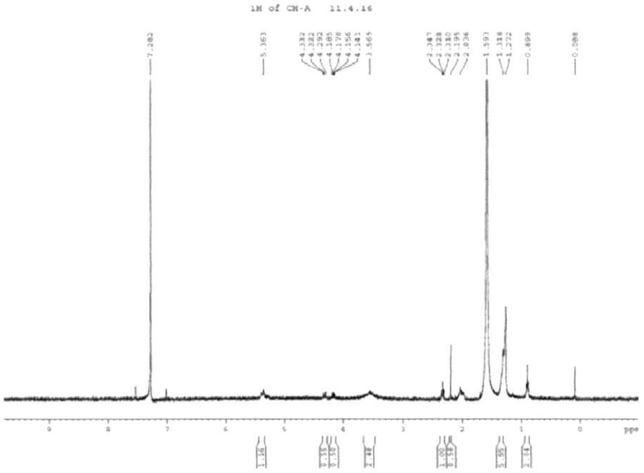

Fig. 7 (a) ^{1}H NMR spectrum of pure SPI

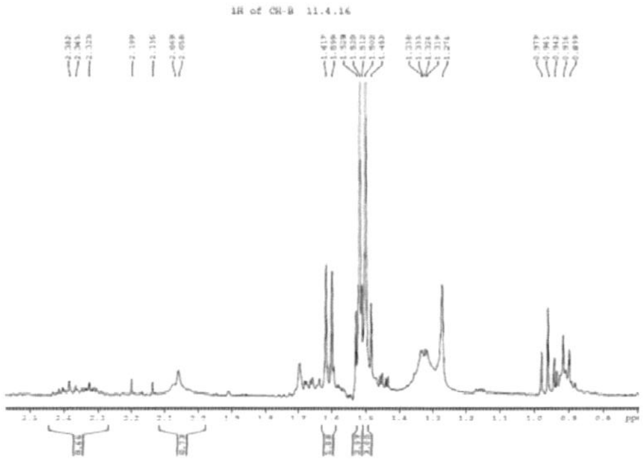

Fig. 7(b) ^{1}H NMR spectrum of Polylactide-g-SPI

Thermal Analysis

Thermo Gravimetric Analysis (TGA):

Primary thermograms of soy protein isolate and Polylactide-g-SPI are presented in Fig. 8 (a and b), respectively. From the figures, it is observed that SPI is stable upto 290°Cwhereas, PLA-g-SPI is stable upto 147°C. Thus, it can be inferred that graft copolymerization of PLA onto SPI has decreased the thermal stability. The higher IDT of SPI is to its granular structure with long protein chains compactly arranged via alpha-helix and beta-sheet. However, the branched structure of PLA-g-SPI destroyed the regularity of the macromolecule chains to a certain extent, which may have unfolded the SPI backbone, therefore reducing the thermal stability. Further, Initial decomposition temperature (IDT), final decomposition temperature (FDT), and decomposition temperatures (DT) at every 10% weight loss are presented in Table 2. The Percent residue left after 600°C for

SPI is 3.4%. The IDT and FDT of polylactide-g-soy protein isolate are 146.94°C and 541°C. The decomposition temperature at every 10 % weight loss for the grafted SPI is observed to be less than that for neat SPI which clearly indicates that the grafted sample is less thermally stable than the neat SPI. However, the percent residue left for the grafted sample is 5.5% which is higher than that of SPI, which came from the ash decomposed from the protein at high temperature. Thus, slightly higher FDT values and percent residue indicate that grafting of polylactide onto soy protein isolate provides comparable thermal stability to SPI.

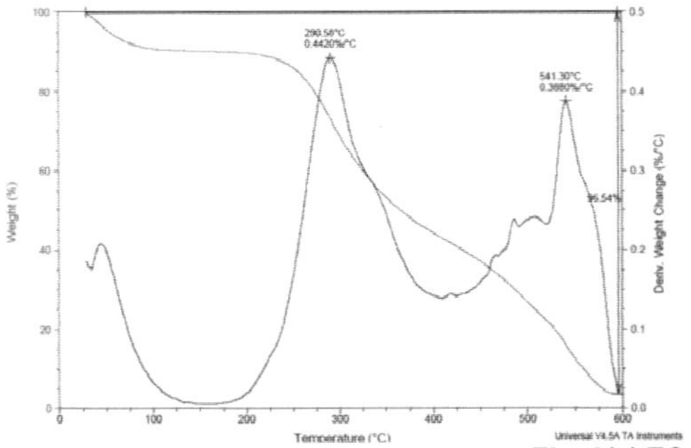

Fig. 8(a) TGA of neat SPI

There were two obvious peaks in the differential thermal gravimetric (DTG) curves of SPI and three peaks for polylactide-g-SPI as can be seen from Figure 8 (a and b) respectively, showing that the thermal degradation could be divided into two and three stages respectively.[5] The first stage for SPI was assigned to the thermal degradation of SPI occurring due to the cleavage of the covalent peptide bonds. During this period, three simultaneous changes occurred: the 7S and 11S protein subunits became dissociated, the secondary structure unfolded, and the denatured subunits reassociated with disulfide, hydrophobic, and electrostatic bonds, and so on. In the second stage, further heating caused breakage of S-S, O-N, and O-O linkages in the structure of soy protein, and at last, the protein molecules completely decomposed and formed various gases such as CO, CO_2, NH_3, H_2S, and others.[19] However, in the DTG curve of polylactide-g-SPI, the first stage is corresponding to thermal degradation of grafted polylactide chains, the second stage corresponds to cleavage of the covalent peptide bonds and third stage corresponding to decomposition of S-S, O-N, and O-O linkages in the structure of SPI and polylactide molecules.[5]

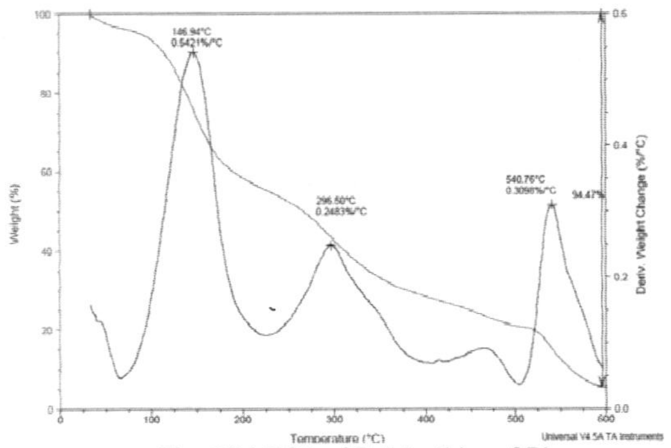

Fig. 8(b) TGA of Polylactide-g-SPI

Differential Scanning Calorimetry (DSC):

Further, the DSC plots of SPI and PLA-g-SPI are shown in Figure 10 (a) and (b) respectively. It has been observed from the Figure 9 (a) that SPI shows broad endothermic peak at 52-165°C, whereas, PLA-g-SPI shows two different endothermic peaks. The first endothermic peak at 85.51°C corresponding to the decomposition of SPI and the second peak at 190°C is corresponding to the melting point of PLA. The appearance of these two peaks is at higher temperature than the peak observed for SPI which indicates that % crystallinity of PLA-g-SPI is increasing upon graft copolymerization and is more stable than neat SPI.

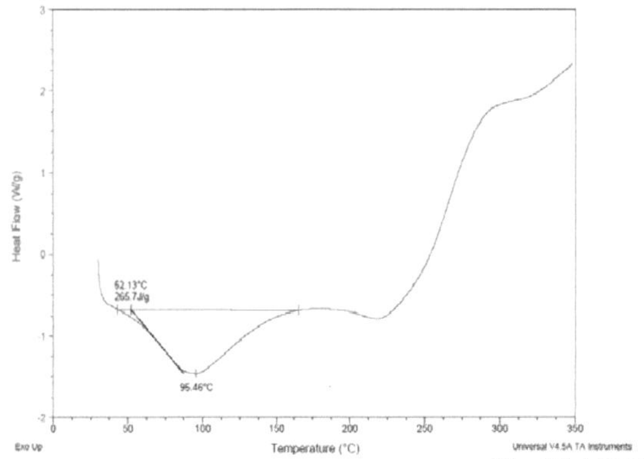

Fig. 9 (a) DSC of SPI

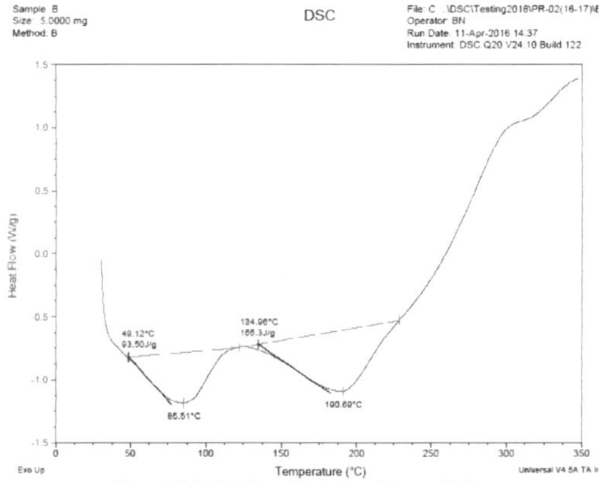

Fig. 9(b) DSC of Polylactide-g-SPI

Compared to the ungrafted SPI, the grafted SPI shows a glass transition temperature (T_g) of PLA at 49°C and melting point of PLA at 190°C as can be seen from Figure 9(b).[18] This transition coincided with the melting point of polylactide homopolymer and would thus be associated with the movement of polylactide side chains, as PLA homopolymer was removed by extraction. Thus, graft copolymerization of PLA onto SPI is confirmed.[5]

X-RAY DIFFRACTION (XRD) ANALYSIS

The powder wide angle X-ray diffraction (XRD) pattern of SPI and Polylactide-g-SPI samples was used to determine the structure for the samples of ungrafted and grafted SPI. As it was observed that the diffraction patterns of the SPI samples exhibited a dominant amorphous halo, a broad band with a maximum at 2θ = 20°, which is characteristic for pure SPI which has 7S and 11S amorphous globulins as the main components.[20]

However, the graft copolymerization of polylactide onto SPI has increased the crystallinity as can be seen from the area under the crystalline region. The XRD pattern for PLA exhibits characteristic peaks at 2 theta values around 15°, 16°, 18.5° and 22.5° due to the presence of L-isomer and peaks at 12°, 21° and 24° due to the presence of D-isomer.[21,22] But WAXD pattern of the grafted sample did not show any sharp peak. This may be due to the fact that the peak for SPI is a broad peak around 20° and the peaks due to polylactide are also around 20° which overlap being in the same region and thereby suppressing the sharp peaks.[23]

When a polymer is in the amorphous state, the macromolecular end groups are distributed randomly in the whole polymer matrix while in a semicrystal polymer, the macromolecular end groups are mostly distributed in the amorphous phase and mesophase. Therefore, when graft copolymerization takes place, the macromolecular end groups are redistributed and result in a rather high concentration of macromolecular end groups in the amorphous phase and mesophase and thereby increases % crystallinity.[24] Therefore, percentage crystallinity of grafted sample is found to increase in comparison to virgin SPI. It is well known that degree of crystallinity is one of the major parameter responsible for tensile strength. Increase in the percentage of crystallinity increases tensile strength. This attributed to the fact that tensile strength of grafted sample polylactide-g-SPI is more than neat SPI.

Field Emission Scanning Electron Microscopy (FESEM) Study

Figures 10 (a) and (b) represents the scanning electron micrographs of SPI and Polylactide-g-SPI, respectively. The comparison of the scanning electron micrographs of the grafted sample with that of ungrafted sample gives a clear indication of change in the topology of the grafted sample. Grafting of polylactide onto Soy protein isolate opens up its matrix and shows considerable deposition of polylactide on the surface of backbone polymer soy protein isolate. It can be observed from Fig 10 (a) that the virgin SPI is neither uniform in size nor uniform in shape. A larger number of reactive sites uniformly created on the SPI molecules led to greater accessibility of the Polylactide and improved considerably the grafting efficiency in the grafting conditions using the swelled SPI.[4,25] It is evident from Fig. 10 (a) and (b) that the SPI matrix opens up and Polylactide molecules are embedded on the surface of SPI molecules. Thus, on the whole, SEM micrographs have provided substantial morphological evidences in favor of grafting of Polylactide onto SPI.[8]

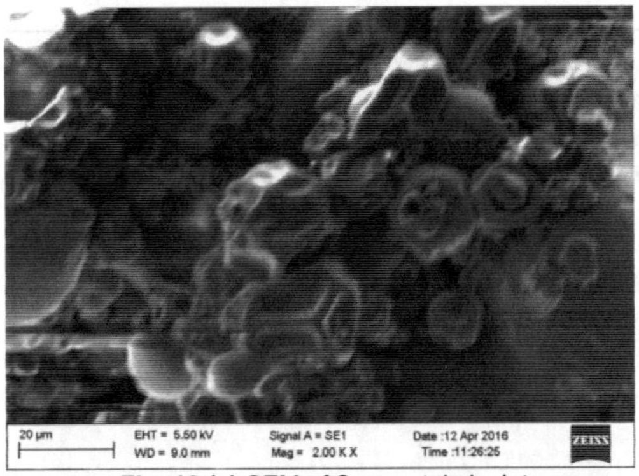

Fig. 10 (a) SEM of Soy protein isolate

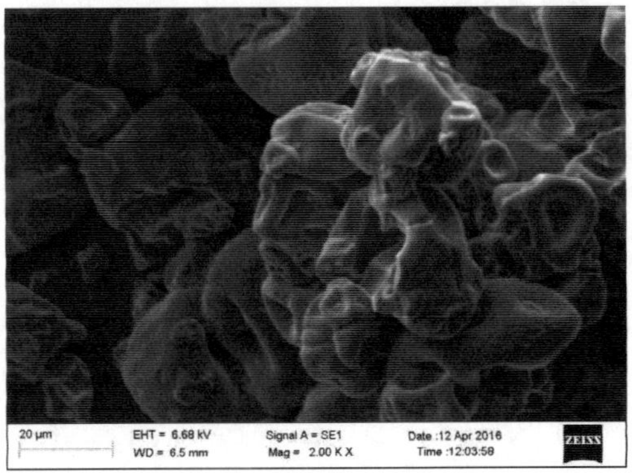

Fig. 10(b) SEM of Polylactide-g-SPI

Swelling and Solubility Behavior of Polylactide-g-SPI

Swelling behavior of SPI and Polylactide-g-SPI in polar solvents was studied in triplicate. The average of the triplicate reading was considered. Samples each of SPI and Polylactide -g-SPI were suspended separately in 10 ml of the ethanol and water in a different proportion. The percent swelling and percent solubility were determined as described in Experimental Section and the results are depicted in Table 2. It is clear that percent swelling of soy protein isolate increases with increasing amount of water in the water-ethanol solvent system both after 24 h and 48 h. Maximum swelling percentage of 526 % and 591 % was observed in neat water (10 ml) in 24 h and 48 h respectively. This swelling behaviour of SPI is observed due to the hydrophilic character of SPI and which was basically due to formation hydrogen bond interaction between them.[4,26]

Similarly, for the grafted samples of Polylactide-g-SPI, swelling (%) also increases with increase in water content both at 24 h and 48 h. And further, the swelling (%) of Polylactide-g-SPI is found to be less than neat SPI as expected which is due to the incorporation of hydrophobic polylactide units onto SPI. Maximum swelling (%) of 173 % and 500 % is observed with 10 ml water system at 24h and 48 h respectively. The percent swelling in 5:5water: ethanol is intermediate to only water and only ethanol system and the percent swelling in only ethanol system is much lesser than only water system. This might be attributed to the fact that neither SPI nor polylactide are soluble in ethanol.[26]

Moreover, the solubility of SPI is found to increase with increase in amount of water after 24 h but does not follow a particular trend after 48 h. The maximum solubility observed after 24 h is 54.8 %. The percent solubility with 5: 5 water:ethanol is intermediate (33.33 %) between only water (70.21 %) and only ethanol (64.47 %) solvent system after 48 h. But for the case of Polylactide-g-SPI, solubility (%) is around 52-55 % in all the cases after 24 h and 6-8 % after 48 h. The decrease in solubility (%) with increase in time from 24 to 48 h might be due to unfolding and denaturing of SPI units during the first 24 h and thus losing its capacity to hold water molecules.

A lower value of % swelling and % solubility in case of grafted samples are observed which might be due to the presence of hydrophobic Polylactide units along with hydrophilic soy protein. The solubility of the grafted samples was found to be less than the neat SPI, which might be due to the incorporation of hydrophobic Polylactide units which improved the water resistance. Thus, their mechanical properties will also be higher and thus, materials made from this graft copolymers might be used for food packaging, coating and biomedical materials. Polylactide -g-SPI has shown greater hydrophobicity behaviour than neat SPI. In these determinations, improved water resistance was observed since the graft copolymers were less swell able and less soluble than the neat SPI.[27]

Table 2 %Swelling and %Solubility of Soy protein isolate, Polylactide-*g*- Soy protein isolate in Water: Ethanol system

Sample	Time	Percentage (%)	10:0	5:5	0:10
Soy protein isolate (SPI)	24	Swelling	526	427	314
	24	Solubility	54.8	18.18	7.31
	48	Swelling	591	544	163
	48	Solubility	70.21	33.33	64.47
Polylactide-*g*- SPI	24	Swelling	173.33	138.09	35.78
	24	Solubility	55.55	54.28	52.63
	48	Swelling	500	385.41	82.22
	48	Solubility	7.5	8.33	6.66

Mechanism of Graft copolymerization of Polylactide onto SPI

Polylactide of high molecular weight are exclusively produced by the ring opening polymerization of the corresponding cyclic monomer lactide.[28] It is formed when cyclic monomers reacted with a catalyst i.e Stannous Octoate ($SnOct_2$).[29] The ring opening polymerization of lactide monomers is based on co-ordination insertion ring opening polymerization, since the propagation is thought to proceed by co-ordination of the monomer to the active species and then insertion of the monomer into the metal oxygen bond by rearrangement of the electrons.[7,30] The growing chain remains attached to the metal through an alkoxide bond during the propagation. The reaction is terminated by hydrolysis forming a hydroxyl end group with functional alkoxy substituted initiators and further, macromers with active end groups are produced which will further help in production of high molecular weight polymers during post polymerization reactions.

It is well known that during the ring opening polymerization of lactides, the catalyst causes trans-esterification reactions at elevated temperatures[31] or at long time reactions.[32] Intermolecular trans-esterification reactions modify the sequences of copolylactides and prevent the formation of block copolymers.

Scheme 1 describes the grafting as well as ring opening polymerization of L-Lactide.[33-36] In step-IV,[37] for the $Sn(Oct)_2$-initiated polymerization of cyclic esters, amino group of Soy protein isolate can also serve as co-initiators.[38,39] More recently, polymerization of L-Lactide initiated by natural amino acids at elevated temperatures has also been reported.[6] These observations open new possibilities, e.g., in the synthesis of polypeptide-poly(aliphatic ester) graft copolymers. On the basis of above discussion the Polylactide grafted Soy protein isolate products are prepared.

Conclusion

A novel biodegradable copolymer, PLA-*g*-SPI, has been successfully synthesized via polylactide grafted onto soy protein backbone by ring-opening polymerization in the presence of an azeotropic solvent. Preswelling

IOP Conf. Series: Materials Science and Engineering **410** (2018) 012011 doi:10.1088/1757-899X/410/1/012011

played an important role as it could promote the graft polymerization effectively. The coordination insertion mechanism has been proposed for the graft copolymerization of polylactide onto soy protein isolate. The thermal properties of the copolymer differed from those of the original protein. The protein chains did not act as a nucleating agent in the copolymer, and crystallization ability was promoted with the increase of graft length or graft yield (%). In addition, the thermal stability of protein was reduced as the branch structure destroyed the compact three-dimensional organization in the soy protein.

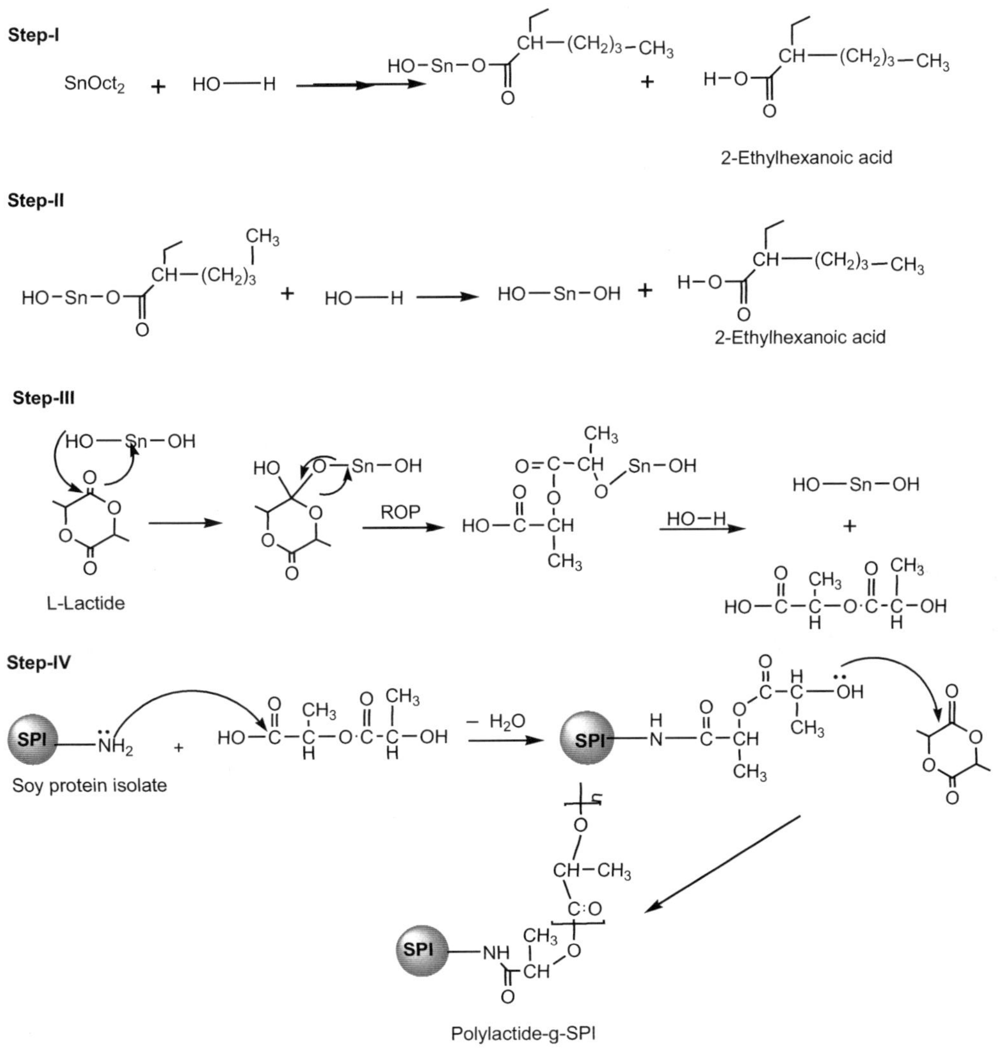

Scheme 1. Reaction mechanism of ROP of Polylactide-g-SPI

Acknowledgements
The authors are grateful to Ravenshaw University and National Institute of Technology, Raipur for providing research facilities.

References

1. Mohanty, A.K.; Misra, M.; Hinrichsen, G. Macromol Mater Eng 2000, 276/277, 1.

2. Chandra, R.; Rustogi, R. Prog Polym Sci 1998, 23, 1273.
3. Thakur, M.K.; Thakur, V.K.; Gupta, R.K.; Pappu, A.; *ACS Sustainable Chem. Eng.* 2016, 4,1.
4. Kaur, I.; Gautam, N. Malaysian Polym J 2010, 5, 39.
5. Li, Y.-D.; Chen, S.-C.; Zeng, J.-B.; Wang, Y.-Z. Ind Eng Chem Res 2008, 47, 8233.
6. Liu, J.; Liu, L. Macromolecules 2004, 37, 2674.
7. Löfgren, A.; Albertsson, A.-C.; Dubois, Ph.; Jérôme, R. J Macromol Sci Rev Macromol Chem Phys 1995, C35, 379.
8. Khanna, N. D.; Kaur, I.; Bhalla, T. C.; Gautam, N. J Appl Polym Sci 2010, 118, 1476.
9. Fanta, G. F.; Trimnell, D.; Salch, J. H. J Appl Polym Sci 1993, 49, 1678.
10. Maharana, T.; Singh, B. C. J Appl Polym Sci 2006, 100, 3229.
11. Patra, C.M.; Singh, B. J Appl Polym Sci 1994, 52, 1557.
12. Ikhuoria, E. U.; Folayan, A. S.; Okieimen, F. E. Int J Biotechnol Mol Biol Res 2010, 1, 010.
13. Xi, D.; Yang, C.; Liu, X.; Chen, M.; Sun, C.; Xu, Y. J Appl Polym Sci 2005, 98, 1457.
14. Sangramsingh, N.M.; Patra, B .N.; Patra, C. M.; Singh, B. C. J Appl Polym Sci 2004, 91, 981.
15. Boua-in, K.; Chaiyut, N.; Ksapabutra, B. *J Optoelectron Adv M* . **2010**, *4*(9); 1404 – 1407.
16. Liu, L.; Wang, Y.; Shen, X.; Fang, Y. *Biopolymers* **2005**, *78*, 163–170.
17. Garlotta, D. *J. Polym. Environ.* **2001**, *9*, 63-84.
18. Ahmed, J.; Zhang, J. X.; Song, Z.; Varshney, S. K. J Therm Anal Calorim 2009, 95, 957.
19. S.N. Swain, K.K. Rao, P.L. Nayak, Polym. Int. 54 (2005) 739-743.
20. Pedro Guerrero Manso, Processing and characterization of soy protein-based materials, PhD thesis 2012.
21. Sarasua, J.R.; Prud'homme, R.E.; Wisniewski, M.; Le Borgne, A.; Spassky, N. Macromolecules 1998, 31, 3895.
22. Ikada, Y.; Jamshidi, K.; Tsuji, H.; Hyon, S.H. Macromolecules 1987, 20, 904.
23. Ahmadi, S.; Morshedian, J.; Hashemi1, S. A.; Carreau, P. J.; Leelapornpisit, W. *Iran* Polym J 2010, 19 , 229.
24. Xu, L.; Dean, K.; Li, L. Prog Polym Sci 2006, 31, 576.
25. Sedighian, A.; Moghbeli, M.R. Starch – Stärke 2010, 62, 49.
26. Zhong, Z.; Sun, X. Polymer 2001, 42, 6961.
27. González, A.; Igarzabal, C.I.A. *Food Hydrocolloids* 2013, *33*, 289.
28. Sobczak, M. Polym Bull 2012, 68, 2219.
29. Selukar, B. S.; Parwe, S. P.; Mohite, K. K.; Garnaik, B. Adv. Mat. Lett. 2012, 3, 161.
30. Mecerreyes, D.; Jérôme, R.; Dubois, P. Adv Polym Sci 1999, 147, 1.
31. Duda, A.; Florjanczyk, Z.; Hofman, A.; Slomkowski, S.; Penczek, S. Macromolecules 1990, 23, 1640.
32. Bero, M.; Czapla, B.; Dobrzynski, P.; Janeczek, H.; Kasperczyk, J.; Macromol Chem Phys 1999, 200, 911.
33. Storey, R. F.; Sherman, J.W. Macromolecules 2002, 35, 1504.
34. Kricheldorf, H. R.; Kreiser-Saunders, I.; Boettcher, C. Polymer 1995, 36, 1253.
35. Kricheldorf, H.R.; Kreiser-Saunders, I.; Stricker, A. *Macromolecules* **2000,** *33*(3),702–709.
36. Kowalski, A.; Duda, A.; Penczek, S. Macromolecules 1998, 31, 2114.
37. Kowalski, A.; Libiszowski, J.; Biela, T.; Cypryk, M.; Duda, A.; Penczek, S. Macromolecule 2005, 38, 8170.
38. Kowalski. A.; Duda. A.; Penczek, S. *Macromolecules* **2000**, *33*(3), 689–695.
39. Cai, Q.; Zhao, Y.; Bei, J.; Xi, F.; Wang, S. Biomacromolecules 2003, 4, 828.

Development of Novel Constructional Material from Industrial Solid Waste as Geopolymer

Muktikanta Panigrahi[1*], Saralasrita Mohanty[2], Radha Raman Dash[1], Ratan Indu Ganguly[1]

[1]Metallurgical Engineering Department, Gandhi Institute of Engineering and Technology (GIET), Gunupur, Odisha, India
[2]School of Physical Sciences, National Institute of Science Education and Research, HBNI, Jatni, India

Abstract

In present investigations, Geopolymer is made from thermal power plant waste obtained such as Pond Ash. The mechanism involved is that the silicon and aluminium present in the waste reacts with alkali liquid, forming Geopolymer which binds other non-reactive materials in the waste. The highest strength level achieved for as-prepared Geopolymers 19 MPa by curing 1 day. The strength level of Geopolymeric products are 22 MPa (Geopolymeric Motar) and 31 MPa (Geopolymer concrete), respectively. SEM micrographs reveal that there is a gradual transformation from irregular spherical shape to compacted mass which is due to polymeric transformation with increase in curing time which also corroborate with mechanical properties such as compressive strength. DSC isotherms show oozing out of inbuilt water which has accumulated during condensation polymerization reaction. The strength level achieved for optimum combination of variable is found to be comparable to that of standard motar of grade (M15) as is used for constructional purpose.

Key words: Geopolymer, Pond ash, Cement and Compressive strength

1.0. INTRODUCTION

In the recent time, fly Ash based Geopolymer (GP) has emerged as promising new cement-like materials which is alternative in the field of building and construction materials [1-3]. Geopolymers show many superior properties such as high compressive strength, low creep, good acid resistance and low shrinkage [3] compared with OPC based constructional materials. The main ingredient of Geopolymers is Fly Ash which is obtained from coal burning thermal power stations and is thus very beneficial in terms of environmental impacts. Fly Ash/Bottom Ash/ Pond Ash based Geopolymer gives emphasis in reducing carbon dioxide emission [1]. Since Pond Ash is the waste products, this research can lead to the awareness of sustainable development to the society. This is very advisable in sustainable developments to reduce carbon dioxide. At present time, there is limited information on the influence of parameters on Geopolymer available, especially Geopolymer with bottom ash as the fine aggregate. As a result, study on the effect of different parameters on the Fly Ash based Geopolymer with bottom ash as sand replacement is needed. Hardjito and Rangan demonstrated that the benefit of Geopolymer concrete which is cheaper than Portland cement concrete [2,3]. Fly ash based Geopolymer products are prepared as like Geopolymer pastes [4], mortars [5] and concretes [2,3]. The inclusion of naturally occurring materials such as quartz, basalt, granite, sandstone and limestone as aggregates to Geopolymer paste is not only economically favorable but also reduces pore density, reduces crack formation and improves durability [6]. In ordinary Portland cement (OPC) concrete and mortar there is a likelihood of an alkali-aggregate reaction between reactive aggregate and alkalis present in cement ($Na_2O + K_2O$) and $Ca(OH)_2$. The product of the alkali-aggregate reaction results in expansion and subsequent crumbling of the mortar and concretes. [7] found that the calcium in the OPC mortar plays an essential role in the expansive nature of the gels. Fly ash based Geopolymer mortars are less susceptible to alkali-aggregate reaction because the lower calcium content in these systems results in a reaction product that is not expansive [6,7]. Previous researchers have described addition of aggregate to Geopolymer with binder: aggregate at constant ratio 0.5 [5].

The present research reports on the effects of sieving size Pond Ash, curing time and curing temperature on the Geopolymerization process within the binder phase and subsequent mechanical properties of the Geopolymer. Also, the work has done for morphological transformation from as-received Pond Ash to Geopolymeric products. DSC analysis is also done.

2.0. EXPERIMENTAL DETAILS

2.1. Chemicals and Materials

The variety of chemicals and materials such as Pond Ash, sodium silicate, sodium hydroxide, Sika are required for the preparation of GP products. Bulk quantity of Pond Ash is collected from NALCO Navratna Company (Damanjodi, Odisha) and the different percentage of composition is mentioned in **Table 1.**

Table 1. Chemical composition of **Pond Ash** (PA)

Table 1. Chemical composition of **Pond Ash** (PA) [8]

Raw materials	SiO_2	Al_2O_3	CaO	MgO	Fe_2O_3	TiO_2	Cr_2O_3	MnO	P_2O_5	C	LOI
Pond Ash	**62.8**	**28.3**	**0.7**	**0.58**	**3.85**	**1.84**	**0.04**	**0.03**	**0.32**	**1.15**	**0.5**

2.2. Cured as-prepared Geopolymer

Pond Ash based Geopolymer is prepared by taking in the amount of as-received Pond Ash, sodium silicate, sodium hydroxide, and water soluble plasticizers. The process is carried out in following three steps such as Grinding, Mixing and Ramming followed by curing using hot air oven [9].

Step 1: The as received Pond Ash from NALCO Damanjodi is grinded by ball mill followed by sieving through different mesh size such as 240 meshes, 200 meshes and 150 meshes.

Step 2: Now, we are taken 240 meshes of Pond Ash for the preparation of Geopolymeric product. The sieved 240 meshes of Pond Ash is mixed with NaOH, Na_2SiO_3, small amount water and water soluble plasticizers (Sika) in appropriate amount and gel-like material is formed. The different compositions of the mixtures are shown in **Table 2**.

Step 3: Appropriate amount of gel-like materials are put in REMI system. REMI mould is made from iron in cylindrical form, which have 50 mm diameter by 70 mm height. The samples are RAMMING twenty (20) times. It is left 3 minutes for compacting and then de-molded it. It is called as Green samples.

The casted green samples are curried at 70 °C for 24 h in oven. The cured samples wrapped within plastic zipper bags to prevent moisture loss and then stored until the samples are tested. In this way, varieties of Geopolymeric samples are prepared by considering the different parameters such as temperature, time and mesh size.

2.3. CHARACTERIZATION TECHNIQUES

The major testing of as prepared Geopolymeric product is compression testing. It is done in compression testing machine (AIMIL COMPTEST 2000, India). Total maximum load are recorded at the point of fracture. It is followed ASTM standard. Morphological analysis of 70 °C cured Geopolymeric products with as-received Pond ash is carried out by Field Emission Scanning Electron Microscopy (FESEM) with Energy Dispersive X-ray. It is conducted using JEOL, Carl Zeiss Supra 40 field emission scanning electron microscope. Before FESEM experiment, gold coating is done using. Operating voltage is maintained at 30 kV. Differential scanning calorimetry (DSC) of 70 °C cured as-prepared samples is performed using a Perkin Elmer Pyris Diamond analyzer at a heating rate of 10 °C/min in nitrogen environment.

2.4. RESULTS AND DISCUSSION

The results of compressive strength of prepared Pond Ash based Geopolymeric sample after different length of curing time at 70 °C are summarized in **Table 2**.

Table 2 Variation of curing time with other parameters as curing temperature (70 C), 240 mesh size, 12% SS and 3% SH

Sample code	Curing Time (hour)	Pond Ash	Sod. Silicate	Alkali (8 M, NaOH)	Water soluble Plasticizer (Sika)	Compressive Strength (MPa)
S1	4	85%	12%	3%	1-2ml	10.6
S2	8	85%	12%	3%	1-2ml	13
S3	16	85%	12%	3%	1-2ml	15
S4	20	85%	12%	3%	1-2ml	18
S5	24	85%	12%	3%	1-2ml	19
S6	168	85%	12%	3%	1-2ml	20.3

In case of prepared Pond Ash based Geopolymer for 7 days (168 h) curing, the compressive strength is found to be 20.3 MPa; whereas the compressive strength of PA based GP curing at 70 °C for 24 h is found to be 19 MPa. There is slight difference in compressive strength which is better than Fly Ash based Geopolymer [10] and M15 grade concrete [11]. This may believe that it is happened due to the increase in Si/Al ratio and with increasing in NaOH (alkali) %. The compressive strength value is higher in samples with higher Si/Al ratio and high alkali % (i.e., Si/Al ratio 3 and alkali % 8) [11].

Table 3: **Mix proportions of Pond Ash based Geopolymers at 24 h in different temperature**

Sample code	Curing Temp.(°C)	Pond Ash	Sod. Silicate	Alkali (8 M, NaOH)	Water soluble Plasticizer (Sika)	Compressive Strength (MPa)
S1	50	85%	12%	3%	1-2ml	15
S2	60	85%	12%	3%	1-2ml	16.5
S3	70	85%	12%	3%	1-2ml	19
S4	80	85%	12%	3%	1-2ml	15

In addition, the compressive strength of as-prepared Pond Ash based GP product in different curing temperature such as 50 °C, 60 °C, 70 ° and 80 °C, respectively is presented in Table 2. It is clearly indicated that as the curing time increases, the compressive strength increases at 70 °C curing temperature. The highest compressive strength is obtained at 70 °C in 24 h curing time, whereas the lowest compressive strength is estimated at 50 °C and 80 °C in 24 h curing. The corresponding highest value of compressive is 19 MPa. The reason may be due to the thermal energy. It may help to the progression of chain and that gives better strength [12,13].

Table 4 Variation of mesh size with other parameters as curing time 24 h, curing temperature 70 C, 12% SS and 3% SH

Sample code	Mesh size (micron size)	Pond Ash	Sod. Silicate	Alkali (8 M, NaOH)	Water soluble Plasticizer (Sika)	Compressive Strength (MPa)
S1	150	85%	12%	3%	1-2ml	15
S2	200	85%	12%	3%	1-2ml	16.5
S3	240	85%	12%	3%	1-2ml	19

Table 4 shows different particle size of Pond Ash which is obtained from sieving of different mesh size. During the sieving, the mesh size is 240 meshes, 200 meshes and 150 meshes. Pond Ash can be taken same percentages during the preparation of Geopolymer. The highest compressive strength is estimated to be 19 MPa and lower one is 15 MPa. It happens may be due to the effect of densities and pores present in the prepared product [14].

For the morphological analysis of the as-received Pond Ash and cured Pond Ash based Geopolymers is shown in Fig 2. For this analysis, the powdered sample is used. FESEM image of as-received Pond Ash sample is shown in Fig 2A. It is observed that the arrangement of the irregular particles is relatively loose with high porosity and small particle size; whereas prepared Pond Ash based

Fig 1. FESEM images of (A) as-received Pond Ash, PA based GP cured at 70^{0}C for 24 h with 240 meshes (B)

Geopolymer (**Fig 1B**) is compacted mass. The rate of compacted mass is formed in different curing time length at particular curing temperature i.e., 70 °C. This may believe that is the formation of Geopolymer because the particles are connected to each other. EDX analysis of the as-received Pond Ash ensures the presence of particular elements in it and is shown in **Fig 2** (Left side).

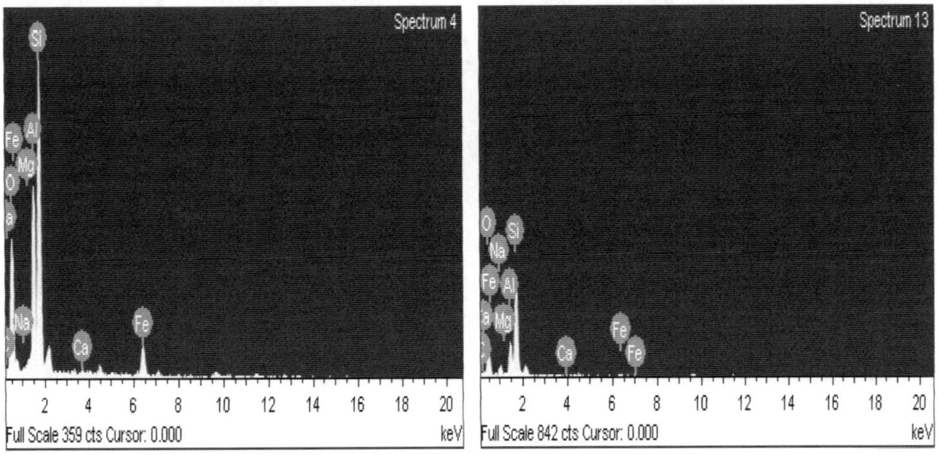

Fig. 2. EDS of as-received Pond Ash (left side) and PA based GP cured at 70 °C for 24 h (right side)

The main elements observed are Si, Al, Fe, Na, Mg, Ca and O. **Fig 2** (Right side) shows the EDS image of as-prepared PA based GP cured at 70 °C for 24 h. The same elements are presented as-like as-received PA. Only the difference is different peak height.

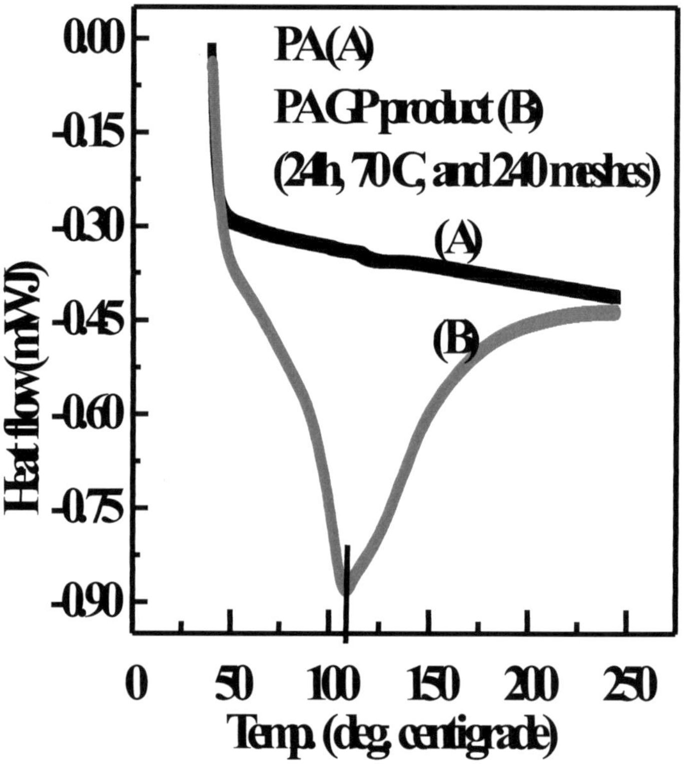

Fig. 3. DSC isotherms of as-received PA (A) and Optimized GP product (B)

Differential scanning calorimetry (DSC) is used to measure a number of characteristic properties of the Geopolymer pastes. From this experiment, it is likely to observe exothermic and endothermic events as well as glass transition temperatures (Tg). DSC isotherm is carried out in the temperature range of50 °C to 300 °C and is shown in **Fig. 3** [15]. In the thermograms, the as-prepared Geopolymer showed exothermic peaks. Similar report is reported [15]. From the **Fig. 3** indicates, the DSC thermogram of Pond Ash is smooth, there is no sign of Geopolymeric reaction. The as-prepared optimized Pond Ash based Geopolymeric product is demonstrated peak in the **Fig. 3**. When the appropriate proportion of Na_2SiO_3, NaOH, SiKA (water soluble plasticizer) and distil water of the optimized product, the exothermal peaks is occurred at a temperature ~ 110 ºC. The sign indicates the Geopolymeric reaction happened. The significant peak is indicated the water's melting point in all Geopolymer sample at 0 ºC [15].

3.0. CONCLUSIONS

Optimization mechanical properties of Geopolymeric process of Pond Ash is carried out with the addition of sodium silicate, sodium hydroxide and water soluble plasticizer with different parameters such as curing temperature, curing time and meshes size. The optimum condition of preparation of Geopolymeric product is 240 meshes, 70 ºC curing temperature with 24 h curing. The optimum strength of developed Geopolymeric product is found to be 19 MPa. In DSC isotherm, the water of crystallization is found to be ~115 ºC. FESEM image of as-received Pond Ash and optimized Geopolymeric product shows the morphological transformation from irregular spherical shape to compacted structure. EDX data indicate the presence of elements such as Si, Ca, Al, Fe, Na, Mg, and O. From the results, it concludes that the construction products such as brick and concrete will be more sustainable in terms of environment performance, economic viability and potential compared to OPC based products.

ACKNOWLEDGEMENTS

The Authors gratefully acknowledge the financial support provided by **Ministry of Mines, Government of India** for carrying out the project work. We also highly acknowledge the **GIET Gunupur** for providing the facility to carry out the works.

REFERENCES

[1] Bilodeau, A., Malhotra Mohan V., "High-Volume Fly Ash System: Concrete Solution for Sustainable Development". ACI Mater. J. 1 (2003) 41.

[2] Krishnan, L., Karthikeyan, S., Nathiya, S., Suganya, K., Geopolymer Concrete an Eco-Friendly Construction Material. Int. J. Research in Eng. and Tech. (IJRET), 03 (2014) 2321.

[3] Allahverdi, A., Kani, E. N., Construction Wastes as Raw Materials for Geopolymer Binders. Int. J. Civil Eng. 7 (2009) 154.

IOP Conf. Series: Materials Science and Engineering **410** (2018) 012012 doi:10.1088/1757-899X/410/1/012012

[4] Ma, H., Cai, J., Lin, Z., Qian, S., Li, V. C., CaCO3 whisker modified Engineered Cementitious Composite with local ingredients. Construct. and Building Mat., 151 (2017) 1.

[5] Sammy M. Nyale. Omotola O. Babajide.Grant D. Birch., Nuran Böke., Leslie F. Petrik Synthesis and Characterization of Coal Fly Ash based Foamed Geopolymer, Procedia Environ. Sci., 18 (2013) 722.

[6] Navid Ranjbar, Mehdi Mehrali, U. Johnson Alengaram, Hendrik Simon Cornelis Metselaar, Mohd Zamin Jumaat, Compressive Strength and Microstructural Analysis of Fly Ash/Palm Oil Fuel Ash Based Geopolymer Mortar under Elevated Temperatures, Construct. and Building Mater. 65 (2014) 114.

[7] Mohd Mustafa Al Bakri Abdullah, H. Kamarudin, Mohammed Bnhussain, W. I. Wan Mastura, Mechanism and Chemical Reaction of Fly Ash Geopolymer Cement- A Review, Asian J. Scientific Research 1 (2011) 247 ·

[8] Kumar, P. H.; Srivastava, A.; Kumar, V.; Singh, V. K., Implementation of Industrial Waste Ferrochrome Slag in Conventional and Low Cement Castables: Effect of Calcined Alumina. J. Asian Ceramic Soc., 2(2014) 371.

[9] Erdem, M.; Altundogan, H. S.; Turan M. D.; Tumen, F. Hexavalent Chromium Removal by Ferrochromium Slag. J. Hazardous Mater. B,126 (2005) 176.

[10] Muktikanta Panigrahi, Pradeep Kumar Rana, Ratan Indu Ganguly, Radha Raman Dash, Structural Transformation of Cured Pond Ash Geopolymers: A Novel Construction Material. In 20th National Conference on Nonferrous Minerals and Metals, 8-9th July 2016; Eds. Rakesh Kumar, K.K.Sahu & Abhilash.

[11] Mishra, Prasad D., Das Kumar S., A Study of Physico-Chemical and Mineralogical Properties of Talcher Coal Fly Ash for Stowing in Underground Coal Mines, Mater. Characterization, 61 (2010) 1252.

[12] Naidu, G., Kumar, P. B. J., Adisesh, S., Satyanarayana, P. V. V., A Study on Strength Properties of Roller Compacted Concrete with 30% Replacement of OPC 53 Grade Cement by Flyash, Int. J. Comput. Engg. Research, 2 (2012) 912.

[13] Vora, R. Prakash, Dave, V. Urmil Parametric Studies on Compressive Strength of Geopolymer Concrete, Procedia Engg,51 (2013) 210.

[14] Hardjito, D., Wallah, S. & Rangan, B., 2002. Study on Engineering Properties of Flyash. J. Australian Ceram. Soc., xxx (2002) 44.

[15] Bakri,M. M A. A., Kamarudin, H., Bnhussain, M., Nizar, I. K., Rafiza, R.A., Zarina, Y., the Processing, Characterization, and Properties of Fly Ash Based Geopolymer Concrete, Rev. Adv. Mater. Sci., 30 (2012) 90.

IOP Conf. Series: Materials Science and Engineering **410** (2018) 012013 doi:10.1088/1757-899X/410/1/012013

ADSORPTIVE REMOVAL OF CD (II) USING ACTIVATED CARBON PREPARED FROM BARK OF FICUSRACEMOSA

***K. Gour[1],D. K. Donadkar[2], P. K. Rahangdale[3]**

[1]*Department of Chemistry, Priyadarshini J.L. College of Engineering, Nagpur*-440009,
[2]*Department of Chemistry Priyadarshini College of Engineering, Nagpur-440 019*
[3]*Department of Chemistry BhawabhutiMahavidyalaya, Amgaon*-441902,

ABSTRACT

Cadmium is toxic to living being and therefore it is essential to remove it from wastewater. Adsorption capacity of activated carbon derived from Ficusracemosabark (ACFRB) for Cd(II) abetment was investigated through batch adsorption experiments. The effects of contact time, pH and adsorbent dose on the removal of Cd(II) have been studied and reported.Adsorption isotherms of Cd (II) on adsorbent were determined and compared with the Langmuir and Fruendlich isotherm models.The percent removal of Cd (II) was found to be increase with adsorbent doses from 1 to 6gm and maximum efficacy was found at 6 gm. At optimum condition nearly 96.47% abatement of Cd(II) has been noted using ACFRB.Maximum adsorption capacity was found at optimum pH 5 and optimum contact time was 110 minutes. The result indicate that the adsorption process is favorable for Langmuir and Fruendlich models.Thus the self-prepared activated carbon under investigation has been proved to be an excellent economical adsorbent material for Cd(II) removal from contaminated water/wastewater.

Key words: Adsorbtion,waste water,adsorbent

1. INTRODUCTION

In the past few years environmental pollution is increased exponentially and reached alarming levels in terms of its effects on living creatures.[1] However environment become polluted not only by rapid industrialization, deforestation and unplanned urbanization, but also some natural phenomenon of anthropogenic activities such weathering of rock and volcanic activities also play a crucial role for enriching the water reservoirs with heavy metals[2,3]. Toxic heavy metals are considered one of the pollutants that have direct effect on man and animals. It is well known that heavy metals are toxic and they can damage nerves, liver and bones, and they can block functional groups of vital enzymes too. Metal ions such as Cd, Cr, Co, Cu, Zn, Pd, Hg, Ni, Ag, and Sr and metalloids such as Se, As, and Sb are toxic if consumed beyond permissible limits.[4,5] Cadmium is one of the most toxic environmental and industrial pollutant because it can damage allmost all important organs.[6] Cadmium and its compounds are also used in paints, pigments, plastics, electroplating, equipments, machineries, baking channels and photography. [7] Even small quantity of Cd assimilation by the body can cause severe high blood pressure, heart disease and can lead to death.[8] The acute over exposure to Cd fumes can cause pulmonary diseases while chronic exposure causes renal tube damage and prostate cancer.[9] The Conventional methods for removing dissolved heavy metal ions include chemical precipitation, chemical oxidation /reduction, filtration, ion exchange, electrochemical treatment and membrane technology. Most of them involve high capital costs with recurring expenses, which are not suitable for smallscale industries. Adsorption method is considered to be one of the preferable methods for the removal of heavy metal ions from aqueous solution due to its

Coresseponding Email: kavitagour1968@rediffmail.com

significant advantages such as low operational cost, wide applicability and creation of relatively low sludge.[10-14] The objective of the present study was to determine Cd(II) ion removal efficiency from aqueous solution using activated carbon derived from the *Ficus racemosa* bark (ACFRB) as an adsorbent; and analysing the data using Langmuir/ Freundlich isotherms.

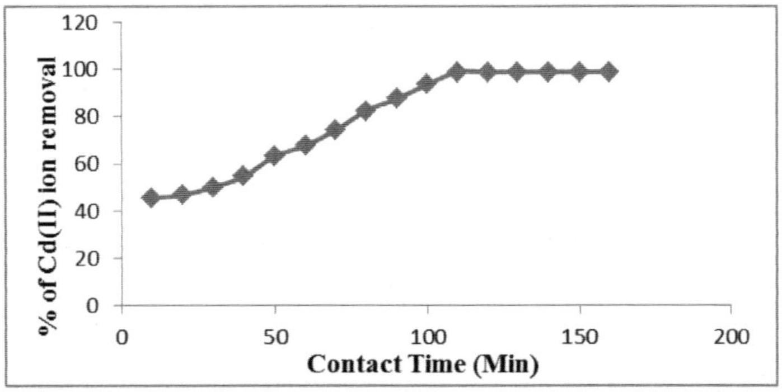

2. MATERIALS AND METHODS

All the chemicals used were of analytical or chemically pure grade. Distilled water was used throughout the investigation.

2.1. Preparation of Activated Carbon from the *Ficus racemosa* bark

Ficus racemosa bark was collected from the nearby local forest area and it was cut into small pieces. It was wash with distilled water and dried in sunlight to remove the moisture. Then it was treated with formaldehyde to avoid the release of color by bark into the aqueous solution during the adsorption process. The above treated bark was carbonized by slow heating over a wide range of temperature (400-700°C) in the absence of air in a muffle furnace. The char obtained was subjected to thermal activation in the absence of air at elevated temperature 900 °C and held at that temperature for 1½ hour. The adsorbent so obtained was ground and sieved through 200 mesh sieves. The dried sample was stored in airtight bottles for further use.[15-17]

2.2. Preparation of stock solution

Stock solution of Cd(II) was prepared by dissolving required quantity of cadmium sulphate salt in the distilled water. This solution was diluted to proper proportions to obtain various standard solutions ranging their concentrations 10-100mgl^{-1} . pH adjustment was done using 0.5N HCl and 0.5N NaOH solution.

3. RESULTS AND DISCUSSION

Effect of contact time on adsorption

For a fixed concentration of heavy metals and a fixed adsorbent mass, the retention of heavy metals increased with increasing contact time before equilibrium is reached. It can be seen that Cd(II) removal efficiency of activated carbon derived from *Ficus racemosa* bark (ACFRB) increased upto 98.77 %, when contact time was increased from 10 min to 110 min. Thus optimum contact time was found to be 110 min.

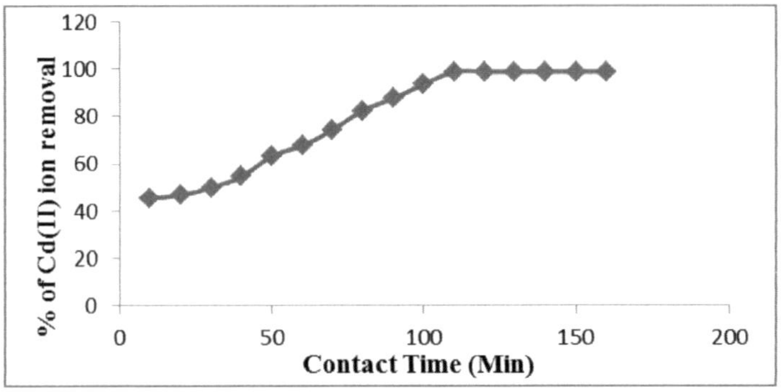

Fig. 1: Effect of contact time on Cd(II) ion removal by activated carbon derived from *Ficus recemesa* bark

Effect of pH on adsorption

Effect of pH on Cd(II) adsorption using activated carbon derived from *Ficus racemosa* bark (ACFRB) as an adsorbent has been studied in the pH range 1to10 and presented in Fig.2. It is seen that solution pH plays a very important role in the adsorption of Cd(II). The percentage removal increases up to 96.27% when pH is increased from 1 to 5 in Cd(II) adsorption and slowly decreases on further increases in pH.

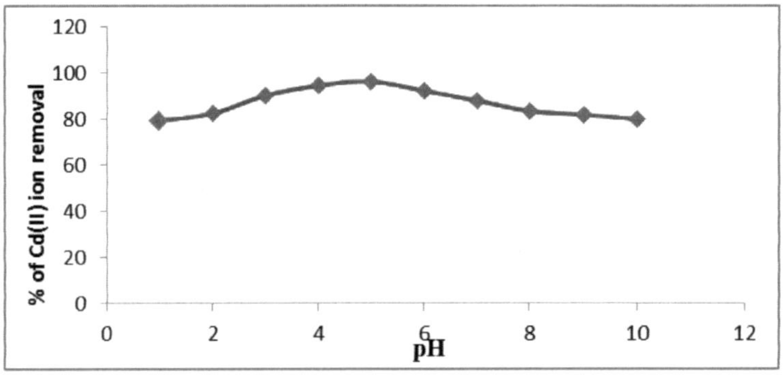

Fig. 2: Effect of pH on Cd(II) ion removal by activated carbon derived from *Ficus racemosa* bark

Effect of Adsorbent Dosage:

The effect of adsorbent doses on percent removal of Cd(II) in the range 1 to 10gm is represented in Fig.3. The obtained results reveal that the percentage removal of Cd(II) ions increased with an increase in the adsorbents dose but after certain adsorbent dose it becomes constant and it is treated as an optimum adsorbent dose. Maximum adsorption was observed at 6 gm/lit i.e. 96.47%.

IOP Conf. Series: Materials Science and Engineering **410** (2018) 012013 doi:10.1088/1757-899X/410/1/012013

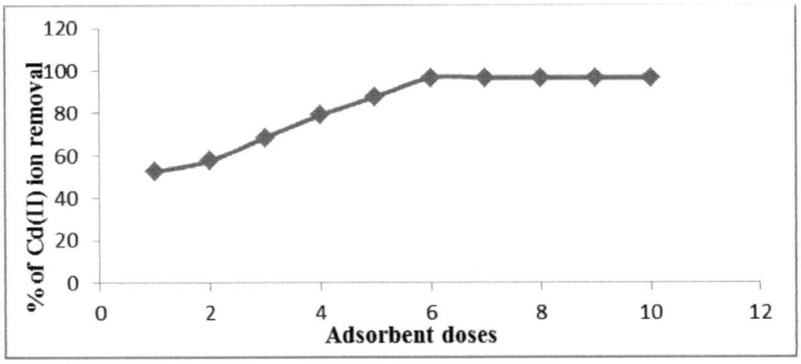

Fig.3: Effect of adsorbent dose on Cd(II) ion removal by activated carbon derived from *Ficus racemosa* bark

Adsorption Isotherm:

Langmuir adsorption Isotherm:

The results obtained from Langmuir model for the removal of Cd(II) by ACFRB has been represented in Fig.4. The values of square of the correlation coefficient (R^2) is found to be 0.9895 for Cd(II), which show the best fitting of equilibrium data. The adsorption efficiency 'Qm' value for Cd(II) was found to be 12.82mg/g while values of 'b' was 0.229. The lower values of b (less than one) implies an excellent the affinity between solute and sorbent sites. To confirm the adsorbility of the adsorption process, the equilibrium parameter also called separation factor 'RL ' for Cd(II) was calculated. The value of RL lies between 0 and 1 for favourable adsorption, while RL> 1 represents favourable adsorption and RL=1 represents linear adsorption while the adsorption process is irreversible if RL=0. The dimensionless parameter RL values lies between 0.466 is consistent with the requirement for favourable adsorption.

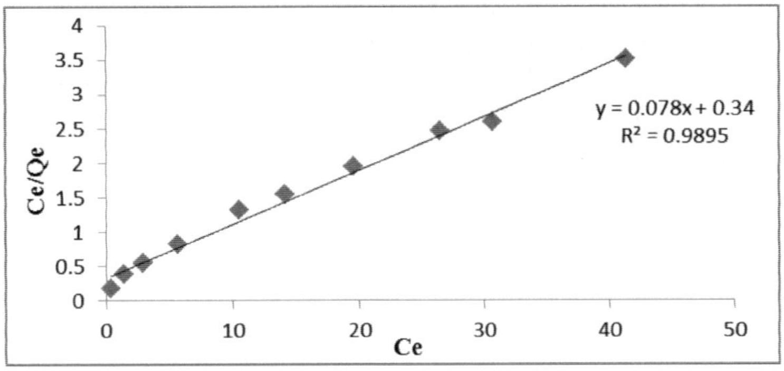

Fig. 4: Langmuir isotherm for the adsorption of Cd(II) ion on activated carbon derived from *Ficus racemosa* bark

Freundlich adsorption isotherm:

The plot of logCe versus log Qe for Cd(II) is presented in Fig.5 which show linear curve and hence the adsorption process obeys Freundlich adsorption isotherms. Freundlich constants 'n' and 'kf' for Cd(II)) were found to be 2.665 and 3.213mg/g respectively. The square of the correlation coefficient (R^2) values was found to be 0.9761for Cd(II) which shows well-fitting of the Freundlich isotherm. The 'n' values are in between 1 to 10 which indicate the favorable adsorption of Cd(II) on ACFRB.

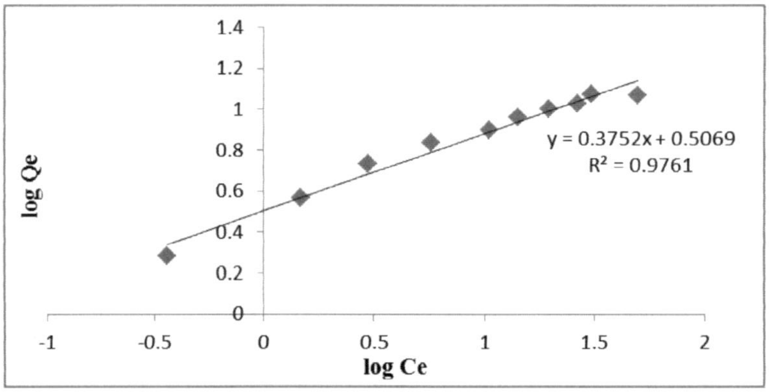

Fig.5: Freundlich isotherm for the adsorption of Cd(II) ion on activated carbon derived from *Ficus racemosa* bark

4. CONCLUSIONS

Activated carbon successfully generated from naturally available raw material that is the *Ficus racemosa* bark, which is abbreviated as "ACFRB". The ACFRB was found to be most effective for Cd(II) ion removal. At pH 5, 96.27% of Cd(II) was removed from aqueous solution and adsorption was found to be pH dependent. Maximum Cd(II) removal is 96.47 % for 6gm/lit optimum adsorbent dose at 110 min of optimum contact time. The experimental data for the adsorption process were well fitted by the Langmuir adsorption isotherm model and Freundlich adsorption model too. Thus the newly generated *Ficus racemosa* bark based activated carbon (ACFRB) reported in this research article has been proved to be an excellent eco-friendly and low-cost adsorbent material which can be successfully used for elimination of Cd(II) from contaminated water.

ACKNOWLEDGEMENTS

Authors are thankful to the Principal and Head of the Chemistry Department, Priyadarshini College of Engineering, Nagpur, India for providing the necessary laboratory and library facilities.

REFERENCES

1. Swathi A Patel "Removal of Heavy Metal Using Natural Adsorbents", 2012, M.Tech Thesis.
2. Y. Jonathan, 2010, American Journal of Applied Science, Vol. 7, pp. 153.
3. B.Y. Kamruzzaman, 2009, American Journal of Applied Science, Vol. 6, pp. 14.
4. S. Arivoli, M. Hema, M. Karuppaiah, and S. Saravanan (2008), E-Journal of

IOP Conf. Series: Materials Science and Engineering **410** (2018) 012013 doi:10.1088/1757-899X/410/1/012013

Chemistry, Vol. 5, pp. 820.

5. K. Mohanty, M. Jha, B.C. Meikap, and M.N. Biswas, 2005, Chemical Engineering Science, Vol. 60, pp. 3049.
6. ATSDR-2008, Draft toxicological profile for cadmium, Agency for toxic substances and Diseases Registry, U. S. Department of Health and Human Service, Atlanta USA.
7. M. Csuros and C. Csuros, 2002, Environmental Sampling and Analysis for metals, London: Lewis Publisher.
8. J. Pan, J.A. Plant, N.Voulvouliis, C.J. Oater, and C.Ihlenfeld, 2010, Environ. Geochem.Health, Vol. 32, pp. 1.
9. L. Jarup and A.Akesson, 2009, Toxicology and Applied Pharmacology, Vol. 238, pp. 201.
10. P.Venkateswarlu, M.V. Ratnam, D.S. Rao and M.V. Rao, International Journal of Physical Sciences, 2007, Vol. 2, pp. 188.
11. A. Shafaghat, F. Salimi, M. Valiei, J. Salehzadeh and M. Shafaghat, African Journal of Biotechnology, 2012, Vol. 11, pp. 852.
12. A.A. Mengistie, T.S. Rao, and A.V. Prasada Rao, Global Journal of Science Frontier Research Chemistry, 2012, Vol. 12 (No. 1), Version 1.0, pp.1.
13. S.L. Pandharipande, U. Dhomane, P. Suryawanshi, and N. Dorlikar, International Journal of Advanced Engineering Research and Studies (IJAERS), 2012, Vol. I, pp 214.
14. S.L. Pandharipade, Y. Moharkar, and R. Thakur, 2012, International Journal of Engineering Research and Applications (IJERA), Vol. 2, pp.1337.
15. D.K.Donadkar, P.K.Rahangdale, and K.Gour, 2016, Der Pharma Chemica, Vol. 8, pp. 230.
16. N.D.Vilayatkar, P.K. Rahangdale, and D.K. Donadkar, 2016, International Journal of Advanced Research, Vol 4, pp. 1360.
17. D.K. Donadkar, P.K. Rahangdale, and K. Gour, 2017, International Journal of Advanced Research, Vol.5, pp. 204.

IOP Conf. Series: Materials Science and Engineering **410** (2018) 012014 doi:10.1088/1757-899X/410/1/012014

A Comparative Investigation on Tribological Properties of Al-Si Alloy based MMC Reinforced with Waste Materials

[1]Debashish Dash, [1] Sandeep Kumar Sahu, [2] Dr. Ajit Kumar Senapati
Mechanical Eng. Department, GIET, Gunupur

Abstract— In the present study Al-Si alloy (LM6) was reinforced with 15% of flyash and rice husk ash. This experimental analysis shows that Al-Si Alloy based MMC reinforced with fly ash and rice husk ash have a better Tribological behaviour as compared to the Al-Si alloy. The wear test was carried out using pin-on-disk machine with a constant load of 30N with varying sliding speed. The experiments are conducted at both elevated and ambient temperature. Based on the observations of weight loss, wear graphs and C.O.F graphs corresponding observations were drawn. With the help of these experiments it was found that the Al-Si alloy reinforced with waste materials shows a better wear behaviour than the existing Al-Si alloy.

Index Terms— Aluminium alloy MMC, reinforcement, tribological properties, wear resistance, waste materials

1. INTRODUCTION

Metal matrix composites (MMCs) create a new era of engineering materials in which a strong reinforcement is added into a metal matrix to improve its wear properties and Tribological properties even at elevated temperature. In the recent years d metal matrix composites (MMCs) have been transformed into a topic of scientific and intellectual interest to a material of broad technological and commercial significance [3,4].Because of their excellent corrosive properties it has various underwater applications. They show excellent properties at high temperature the reason behind their importance in aircraft engines. Because of the light weight and increased performance they are hot picks for sports car chassis. Aluminium silicon based MMCs have received increased attention in recent decades as engineering materials with most of them possess the advantages of high strength and wear resistance [9,10]. Fly ash is the waste material of every coal based power plants. So it costs zero .It has low density so ultimately reduces the weight of the MMC and has some of the important components like SiO_2, Al_2O_3, Fe_2O_3 and CaO. So it can be used as a reinforcing agent. Rice husk is a waste product of the agricultural farms. The rice husk ash can be used as a reinforcing agent to aluminium –silicon based composite. Many experiments have been carried out to test the wear properties at ambient temperature. But very little work has been done at elevated temperature. So this experimental work has been conducted at various elevated temperatures. The results of these experiments show a great improvement to the existing alloys.

2. EXPERIMENTAL PROCEDURE

2.1. Material Composition

In this experiment aluminium silicon alloy (LM6) was used as the matrix containing 12.25% of silicon and 86.73% of aluminium .It also contains many other materials the composition of which is given in the Table 1.

Table 1: Wt% of Aluminium-Silicon Alloy

AL	Si	Fe	Mn	Cr	Zn	Cu
85.73	12.25	0.45	0.18	0.1	0.09	0.09
Ti	Sn	Ni	Co	Ca	V	
0.06	0.05	0.03	0.03	0.02	0.01	

Flyash as reinforcement to Al-Si Alloy

As we know flyash is a low cost reinforcement so it is used in the Al-Si alloy based mmc. The composition of flyash used in the experiment is given in Table 2.

Table 2: Composition of Fly Ash

Compound	Na_2O	CaO	MgO	SiO_2	AL_2O3	K_2O	Fe_2O_3
Wt%	0.10	1.24	0.58	56.43	31.54	0.59	4.97

Rice-Husk Ash as reinforcement to Al-Si Alloy

Rice husk ash is a major reinforcing agent now a day in AL-SI alloy because of its characteristics such as easily availability and low cost. The composition of flyash used in this experiment is given in Table 3.

Table 3: Composition of Rice husk Ash

Compound	Fe_2O_3	SiO_2	CaO	Al_2O_3	MgO	L.O.I
Wt%	0.93	67.33	1.35	4.90	1.79	17.80

2.2. Production of Metal Matrix Composite

The Al-Si ingot was cut into required dimensions after being cleansed thoroughly. Then they were weighed and finally put into a bottom poured melting furnace (Fig. 1). To improve the wet ability of the MMC the fly ash was pre heated to 670C, and then 15% wt. % fly ash was added to the Al-Si alloy. The preheating also helps to remove any residual moisture present and it is in Muffle Furnance (Fig. 2). A coated stainless steel rotor was used to stir the molten metal at 450rm. Due to this a vortex is created with the molten metal where the fly ash which was preheated was poured slowly at the centre of the vortex. To maintain a distance of 11mm the motor was slowed from top to bottom. The pouring temperature of the liquid was maintained at 750c .A rectangular mould of dimension (250*20*45mm, Fig. 4) was prepared for the casting.

Fig.1-Bottom Pouring Furnace

Fig.2-Muhffle Furnace

The same Al-Si ingot was cut into required dimension after being cleaned properly. Before getting the ingot into furnace it is pre heated to 800c then they are put into the bottom pouring furnace for melting. The rice husk ash is preheated to 675c for removing the moisture content from it. The molten Al-Si was stirred at 450 rpm. When the vortex was created due to the rotation the pre heated rice husk ash was added to its centre. The rice husk added is 15 wt. % .Stir casting technique is followed for the proper mixing of the rice husk with the molten Al-Si. At last a rectangular mould of dimension (250*20*45).The Al-Si pre-heater is shown in the Figure 3.

2.3. Testing of Wear Rate at Elevated Temperature

The wear tests were carried out in the pin-on-disc wear testing machine. All the experiments were conducted at elevated temperature conditions with varying loads. The test samples had dimensions of 8mm diameter and 27.5mm in length. The samples were made to slide on the low alloy steel disc (material EN-31 HRS 60 W 61 cqual to 4340).The track diameter was set to 75mm.Before beginning the experiment the track was cleaned with acetone. The experimental parameters including the time and temperature were set as required. The wear and the frictional force are brought to 0 before beginning. The wear and coefficient of friction electric sensors attached to the machine give the reading in form of graphs. Electric sensor weighing machine was used to measure the initial weight before beginning the wear test and after conducting the experiment the samples were cleaned with acetone and again weighed with the weighing machine to get the weight loss.

3. RESULTS AND DISCUSSION

The experiments were conducted using pin-on-disk machine (Fig. 3) and the wear were recorded by the sensors attached to the machine and weight loss was measured using electronic weighing machine. The results were plotted in form of graphs. The test conditions for flyash, rice husk ash and the Al-Si alloy are given in Table no 4, 5, and 6 respectively. The wear graph of the three samples at ambient temperature is given in Fig 5, and Fig. 6 represents the wear results at elevated temperature.

Fig.3- Pin-ON-Disk Wear Machine

Fig-4- Wear Samples

IOP Conf. Series: Materials Science and Engineering **410** (2018) 012014 doi:10.1088/1757-899X/410/1/012014

Table 4-Test Conditions for Al-Si Alloy

SL. No.	Load (N)	Temp.	Time (sec)	Sliding Speed (m/s)	Int. wt. (gm)	Final Wt. (gm)	Wt.Loss (gm)	Avg Wt.Loss	C.O.F	Avg. C.O.F
1	30	150	1000	1	3.521	3.553	0.032	0.0335	0.345	0.392
2	30	150	1000	2	3.534	3.569	0.035		0439	
3	30	Room Temp.	1000	2	3.561	3.588	0.027	0.025	0.298 1	0.2927
4	30	Room Temp.	1000	1	3.555	3.578	0.023		0.287 3	

Table 5- Test Results for Fly Ash

SL. No	Load (N)	Temp .	Time (sec)	Sliding Speed (m/s)	Int. wt. (gm)	Final Wt. (gm)	Wt. Loss (gm)	Avg Wt Loss (gm)	C.O.F	Avg C.O.F
1	30	150	1000	3	3.553	3.548	0.005	0.0035	0.224	0.2485
2	30	150	1000	4	3.546	3.544	0.002		0.273	
3	30	Room Temp .	1000	3	3.557	3.556	0.001	0.0015	0.213	0.2275
4	30	Room Temp	1000	4	3.546	3.544	0.002		0.242	

Table 6-Test Results for Rice Husk Ash

SL. No	Load (N)	Temp .	Time (sec)	Sliding Speed (m/s)	Int. wt.	Final Wt. (gm)	Wt. Loss (gm)	Avg Wt.Loss (gm)	C.O.F	Avg C.O.F
1	30	150	1000	4	3.537	3.531	0.006	0.0055	0.293	0.351
2	30	150	1000	3	3.588	3.583	0.005		0.409	
3	30	Room Temp .	1000	4	3.543	3.498	0.005	0.004	0.297	0.3095
4	30	Room Temp .	1000	3	3.458	3.455	0.003		0.322	

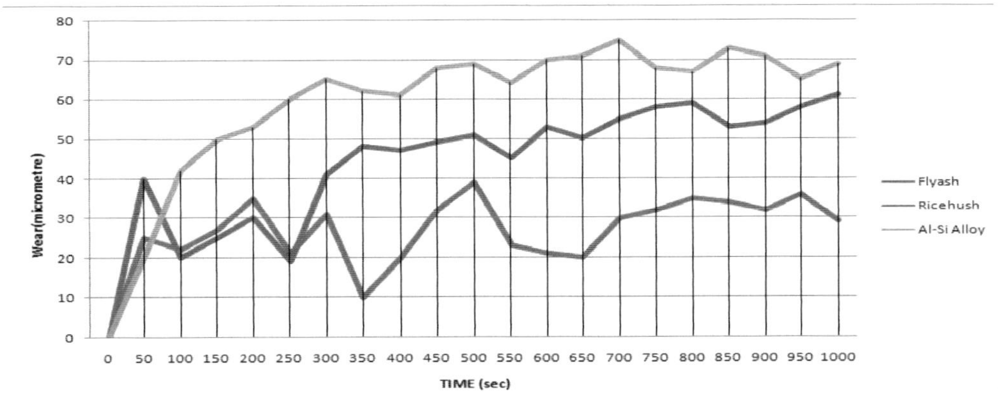

Fig. 5. Representing the wear behaviour of 3 samples at ambient temperature.

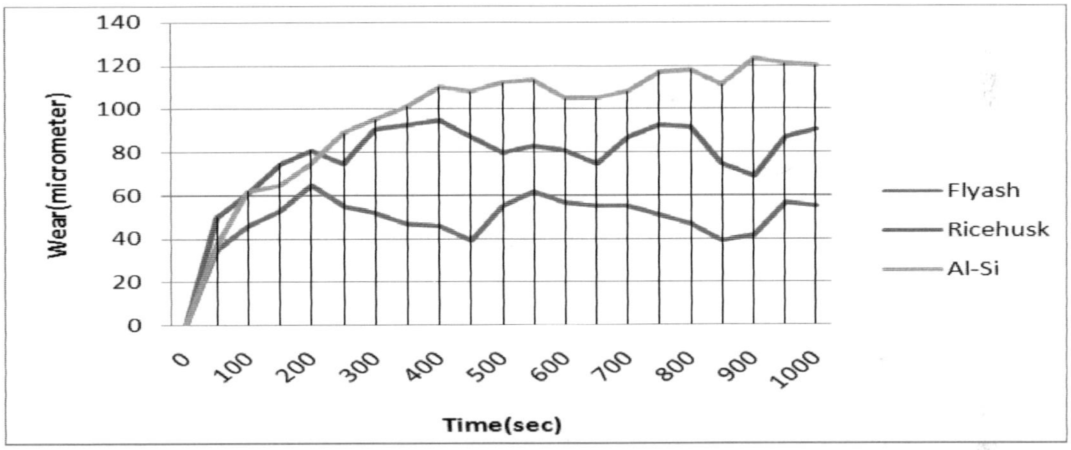

Fig. 6. Representing wear of 3 samples at elevated temperatures.

4. CONCLUSIONS

The stir casting method used for the preparation of composites is easy, efficient and most economical method. It also helps in the uniform distribution of reinforced particles (Fly ash and RHA) with the matrix metal. The wear behavior is higher for MMC with fly ash than MMC with RHA. . Not only at ambient temperature but also at elevated temperatures ,fly ash reinforced MMC showed better results than rice husk reinforced MMC as well as virgin Al-Si Alloy This is because due to the presence of more amount of aluminum oxide and calcium oxide in Fly ash than rice husk ash. The experimented data shows that the selection of reinforced particle when Tribological properties are consider is one of the major aspect for the production of metal matrix composite.

5. REFERENCES

[1] E. A. Feest, 1986, *Materials & Design, Vol.* 7, pp. 58-64.

[2] D. B. Miracle, 2005, *Composites Science and Technology, Vol.* 65, pp. 2526-2540.

[3] M. F. Ashby and D. R. H, Jones, 1980,"Engineering Materials: An Introduction to their properties and applications", International series on materials science and technology; UK: Elsevier Science and Technology.

[4] B.K. Prasad, 2007, Wear, Vol. 262, pp. 262–273.

[5] W. Pedersen and M. Ramulu; 2006, Journal of Materials Processing Technology, Vol. 172, pp. 417–423.

[6] R. D. Day, R.F. Smart, and N. Tommis; Associated Engineering (A.E) Technical Symposium 1970, U.K., Paper No.4.

[7] S. Suresh, A. Mortensen, and A. Needleman, 1993, Fundamentals of Metal Matrix Composites; Butterworth-Heinemann, Stoneham; MA02180, USA.

[8] C. Yan, and L. C. Zhang, 1995, Applied Composite Materials, Vol. 1, pp. 431–447.

[9] O. P. Modi, B. K. Prasad, A. H. Yegnewaran, and M. L. Vaidya, 1992, Materials Science Engineering A, Vol. 151, pp.235-244.

[10] Z. F. Zhang, L. C. Zhang, and Y. W. Mai, 1995, *Journal of Materials Science,* Vol. 30, pp. 5999–6004.

[11] Z. F. Zhang, L. C. Zhang, and Y. W. Mai, 1995, *Journal of Material Science,* Vol. 30, pp. 1961–1966.

[12] F. Toptan, A. Kilicarslan, and I. Kertil, 2010, Materials Science Forum, Vol. 636-637, pp.192-197.

[13] A. Pramanik, L.C. Zhang, and J. A. Arsecularatne, 2006, *International Journal of Machine Tools and Manufacture, Vol.* 46, pp. 1795–1803.

[14] Z. Zhang, J. Zhang, and Y. W. Mai, 1994, Wear, Vol.176, pp. 231-7.

[15] C. Limmaneevichitr and W. Eidhed, 2003, Materials Science and Engineering, Vol. 355A, pp.174–179, 2003.

[15]. M. Babic, A. Vencl, S. Mitrović, I. Bobić, 2009, Tribology Letters, Vol. 36, pp. 125-134.

[16]. M. Babic, S. Mitrovic, B. Jeremic: 2010, Tribology International, Vol. 43, pp. 16-21.

[17] B. Bobic, S. Mitrovic, M. Babic, I. Bobic: 2010, Tribology in Industry, Vol. 32, pp. 3-11.

[18] S. Mitrovic, M. Babic, F. Zivic, I. Bobic, D Dzunic: 2008, Tribology in Industry, Vol. 30, pp. 33- 39.

IOP Conf. Series: Materials Science and Engineering **410** (2018) 012015 doi:10.1088/1757-899X/410/1/012015

The effect of Welding Parameters on Mild Steel by MMAW

Gopal Krushna Mohanta [a], Ajit Kumar Senapati*,

Mechanical EnggDeptt, Gandhi Institute of Engineering Technology, Gunupur,

Rayagda, Odisha, India

ABSTRACT

In this work, effect of welding parameters on the weld ability of mild steel which is welded by Metal arc welding, butt joint is investigated. The current and voltage are chosen as welding parameters. The effect of welding parameters is investigated on depth of penetration and Heat input which are measured as output parameters of welding.

Key words: Mild steel, Arc welding, Welding current, welding voltage, Welding speed

Corroseponding email: *gkmohanty787@gmail.com,senapati.ajit@gmail.com*

1. INTRODUCTION

Today a wide variety of metal joining processes are used in fabrication industries. The welding is majorly used for metal joining. In this process arc i.e. electric discharge between electrode and parent metal is established. Due to high electrical resistance of welding arc high temperature is produced which is enough to melt the metal. MMAW - Manual Metal Arc welding is one of the oldest arc welding processes. Mild steel is widely applicable in fabrication of structure, process equipment, piping and ship building. The major factors affecting mechanical properties of weld joint are welding current, arc voltage, welding speed, polarity, edge preparation and welding technique. Out of these variables welding current, welding voltage and welding speed are primary variable which controls the fusion, depth of penetration, shape of weld puddle, reinforcement and heat input. Electrode polarity, inclination angle and welding technique are secondary variable which affect on energy absorbed, melting rate of base metal and weld metal. So in the present work. The effect welding current, are voltage on mechanical properties of mild steel elements have been investigated.

2. LITERATURE REVIEW

Manual metal arc welding was first invented in Russia in 1888. It involved a bare metal rod with no flux coating to give a protective gas shield. The development of coated electrodes did not occur until the early 1900s when the Kjellberg process was invented in Sweden and the Quasi-arc method was introduced in the UK [1]. It is worth noting that coated electrodes were slow to be adopted because of their high cost. Welding is a wide process used for fabrication to join two materials .The authors considered welding parameters such as voltage, current and welding speed on depth of penetration. The material was carbon alloy steel(0.14%C) of dimension 75 x 50 x6mm,voltage 18v and current was 250amps.The maximum penetration takes place with minimum speed even by maintain constant voltage and current[2]. Welding is more economical, convenient and less susceptible to failure or corrosion as compared to other joining processes. Owing to the advantages of welding over other joining process, numerous welding processes have been Developed [3]. For manual welding methods, labor costs generally make up the vast majority of the total cost. As a result; many cost-saving measures are focused on minimizing operation time. To do this, welding procedures with high deposition rates can be selected, and weld parameters can be fine-tuned to increase welding speed. Mechanization and automation are often implemented to reduce labor costs, but this frequently increases the cost ofequipment and creates additional setup time. Material costs tend to increase when special properties are necessary, and energy costs normally do not amount to more than several percent of the total welding cost [4]. The primary investigation was done to know effect of welding parameters such as current, arc voltage, welding speed, heat input rate on Heat affected zone. It was observed that there is change in structure of material from austenite into bainite which could improve strength and impact toughness. It was known that by increasing speed the width of HAZ decreases and it is necessary to control welding parameters to get better weld quality bead [5].The author investigated multiple weld qualities of welding input process parameters for obtaining greater weld strength with optimum metal deposition rate in welding of dissimilar metals like stainless steel and Mild steel . The process used for welding is Manual Metal Arc welding and dissimilar metals used are low carbon steel and Stainless steel. Welding speed, voltage, current, electrode angle are taken as controlling variables [6].

From the literature survey it is concluded that the most commonly primary variables in are welding arc welding current, voltage & speed, which would change characteristics of joint

2.1 Welding current: Welding current is an influential variable that control melting rate of electrode, there by the deposition of electrode rate on joint. It also controls the depth of fusion, and penetration

2.2 Welding Arc Voltage: Welding arc voltage is also called are arc voltage, is the electrical potential difference between tip of electrode and the surface of molten weld pool. It effects on shape of fusion zone, welding reinforcement electrode melting rate etc.

2.3Welding Speed: Welding speed is the linear speed at which are moves with respect to plate, along with the weld joint. If speed get increases the heat input to the joint decreases and if speed get decreases, heat input increases.

2.4Heat Input: Heat input rate or energy of arc was an important parameter in welding which can be calculated

Heat input rate=V x I x 60/ v J/mm

Where V= Arc voltage in volts

I= Welding current in ampere

v= welding speed or arc travel speed (mm/min)

3. EXPERIMENTAL METHODOLOGY

3.1EXPERIMENT PROCEDURE

Following steps are followed for experiment

1. First 36 specimens of dimension 70mm x40mm x5mm from mild steel flat.
2. 45° single V edge preparation was made on these specimens.
3. Before welding the specimens are cleaned from dust, oil to avoid impurity in molten metal pool.
4. Welding was made by closed butt joint thus edges of the work piece are suitably prepared.
5. Work pieces are kept in position with respect to each other and welding was performed.

IOP Conf. Series: Materials Science and Engineering **410** (2018) 012015 doi:10.1088/1757-899X/410/1/012015

Fig-1EXPERIMENTAL SETUP

3.2 PARENT METAL USED & ITS CHEMICAL ANALYSIS

Mild steel flat pieces of 70 mm x 40mm x 5 mm thickness. The chemical analysis of mild steel is as shown in Table -1.

Table -1 Chemical Composition of Base Metal Element

ELEMENT	Cr	Ni	C	Mn	S	P	Si	N
% of Wt	0.69	0 .01	0.19	0.8	0.04	0.017	0.4	-

3.3 Electrode used & its Chemical Analysis

3.15 mm diameter and 350 mm long electrode was used for welding. Chemical composition of electrode is shown in Table – 2.

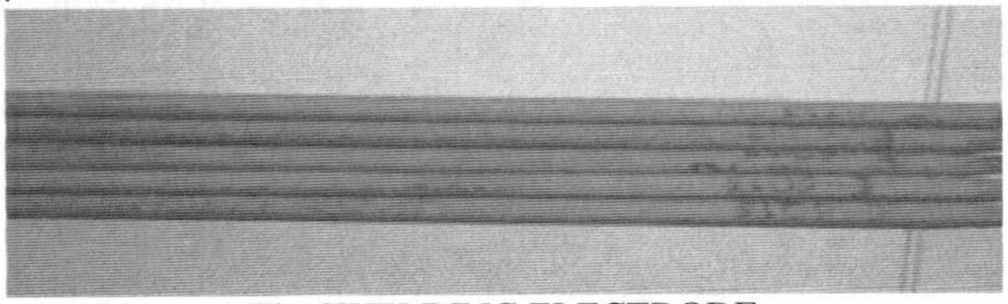

Fig-2WELDING ELECTRODE

Table -2 Chemical Composition of Electrode

ELEMENT	C	Mn	Si	S	P
%of Wt	0.07	0.44	0.22	0.02	0.02

IOP Conf. Series: Materials Science and Engineering **410** (2018) 012015 doi:10.1088/1757-899X/410/1/012015

Fig-3 WELDING MACHINE

In this work metal arc welding is used. Arc welding is a welding process, where Coalescence is produced by heating with an electric arc or arcs, mostly without the application of pressure and with or without the filler metal .The safety precautions & corrective measures taken to avoid accidents and to get good quality weld beds. In welding process speed of welding was varied along with the arc voltage and current. The voltage and current is varied in three steps as shown in following table3.

TABLE 3. VARIOUS PARAMETERS OF WELDING

SLNO	Welding Voltage	Welding Current (Amp)	Welding Speed (mm/min)	Heat Input (J/mm)	Penetration In mm
1	15	205	38.12	2191.94	3.70
2	15	205	37.45	2246.23	3.80
3	15	205	36.91	2258.75	4.52
4	15	205	37.25	2214.02	4.50
5	15	205	36.21.	2205.45	4.30
6	15	205	35.54	2215.23	4.12
7	20	91	21.58	1345.14	3.60
8	20	91	19.87	1249.32	4.20
9	20	91	18.23	1186.21	4.40
10	20	91	18.29	1143.57	4.50
11	20	91	17.58	1097.53	3.09
12	20	91	16.27	983.32	2.80
13	25	122	35.21	1298.02	4.30
14	25	122	30.59	1124.36	4.35
15	25	122	23.62	816.23	2.80
16	25	122	39.12	1423.35	4.60
17	25	122	34.53	1237.23	4.30
18	25	122	30.29	1087.25	3.08

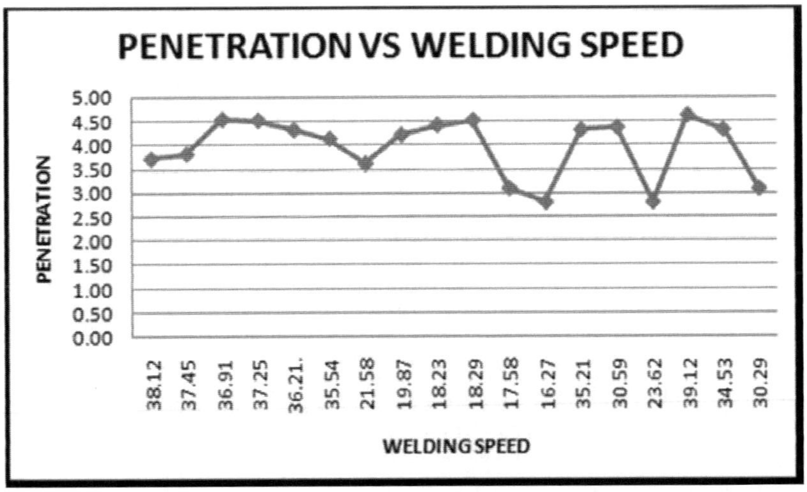
PENETRATION VS WELDING SPEED
PENETRATION
5.00
4.50
4.00
3.50
3.00
2.50
2.00
1.50
1.00
0.50
0.00
38.12 37.45 36.91 37.25 36.21. 35.54 21.58 19.87 18.23 18.29 17.58 16.27 35.21 30.59 23.62 39.12 34.53 30.29
WELDING SPEED

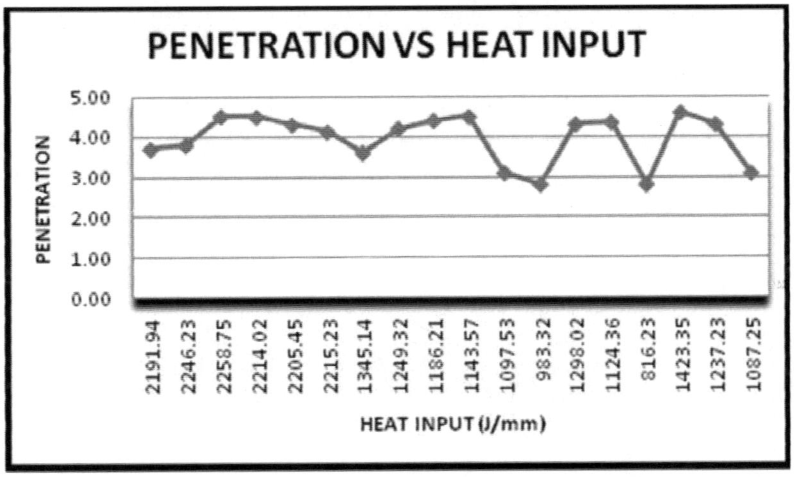
PENETRATION VS HEAT INPUT
PENETRATION
5.00
4.00
3.00
2.00
1.00
0.00
2191.94 2246.23 2258.75 2214.02 2205.45 2215.23 1345.14 1249.32 1186.21 1143.57 1097.53 983.32 1298.02 1124.36 816.23 1423.35 1237.23 1087.25
HEAT INPUT (J/mm)

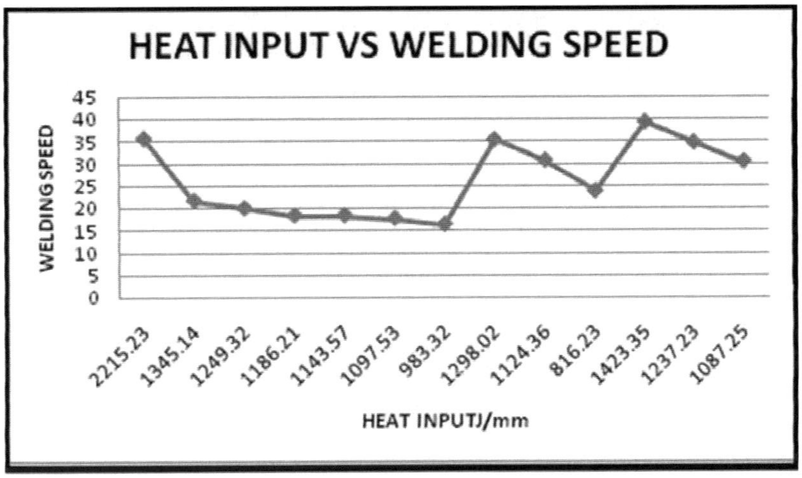
HEAT INPUT VS WELDING SPEED
WELDING SPEED
45
40
35
30
25
20
15
10
5
0
2215.23 1345.14 1249.32 1186.21 1143.57 1097.53 983.32 1298.02 1124.36 816.23 1423.35 1237.23 1087.25
HEAT INPUTJ/mm

IOP Conf. Series: Materials Science and Engineering **410** (2018) 012015 doi:10.1088/1757-899X/410/1/012015

4. RESULTS AND DISCUSSION

The speed of the welding was calculated for each welded specimen and the depth of penetration was measured by cutting cross section perpendicular to direction of welding. The variation of depth of penetration ,welding speed and heat input is analyzed from graph 1,2and 3.From graph it was observed that depth of penetration linearly increases by decreasing the speed and reaches maximum depth and later it get decreases, even though by maintaining constant voltage and current. The depth of penetration is increases at each case of voltages and current with the increased heat input up to maximum and gets decreases gradually as heat input decreases. And least depth of penetration2.8mm was observed at lowest heat input816.23J/mm value even by maintaining constant voltage and current. And maximum depth of penetration is at heat input value 2258.02 J/mm. The graph welding speed Vs Heat input shows that by maintaining constant welding speed a heat input remains constant with small variation .So it can be stated that at different voltage, current and welding speed effects on depth of penetration and heat input.

5. CONCLUSIONS

- Even by maintain constant voltage and current, depth of penetration varies with welding speed and maximum depth was observed at speed of 39.12mm/min for voltage 25Vand current 122amps.
- So it can be concluded that increase in speed and by maintaining constant voltage and current depth of penetration increase up to maximum value and if speed increases beyond the optimum value results to decrease in depth of penetration.

6. REFERENCES

[1] David S A 1986 ASM International
[2] Sharma S K, Mehdi S H I 2013 Int. J.Eng. Adv. Tech. Vol -2
[3] Khanna O. P 1999 Material Science and Metallurgy
[4] Weman K 2003Welding processes. 184-189
[5] Boob AN, Gattani GK 2013 J.Modern.Engg.Research 1493-1500
[6] Patil U S, Kadam M S2013 Int. J. M. Eng.Technolo. 79-85

IOP Conf. Series: Materials Science and Engineering **410** (2018) 012016 doi:10.1088/1757-899X/410/1/012016

Investigation on Mechanical properties of Al-6061 Alloy based MMC

Ajit Kumar Senapati[a], Abhishek-Kumar[b], Anurag-Kumar[b],

[a] Asso.Professor, Mechanical EnggDeptt, GIET, Gunupur,

[b] Final Year student's of Mechanical EnggDeptt, GIET, Gunupur

senapati.ajit@gmail.com, anuragtiwary11@gmail.com

Abstract— In the present study aluminum alloy (6061) composites containing 10 and 15 wt.% of fly ash particles have been fabricated. Some mechanical behavior of unreinforced alloy and metal matrix composites are studied. Charpy impact test and compression test are considered for mechanical properties study. Results show that the Metal matrix composite prepared with 15 wt.% of fly ash exhibit better mechanical property compared to unreinforced alloy as well as metal matrix composite (MMC) prepared with 10 Wt.% of fly ash.

1.INTRODUCTION

Metal matrix composites (MMCs) represent a new generation of engineering materials in which a strong ceramic reinforcement is incorporated into a metal matrix to improve its properties including specific strength, specific stiffness, wear resistance, excellent corrosion resistance and high elastic modulus[1,2]. Over the past two decades metal matrix composites (MMCs) have been transformed from a topic of scientific and intellectual interest to a material of broad technological and commercial significance[3,4]. MMCs combine metallic properties of matrix alloys (ductility and toughness) with ceramic properties of reinforcements (high strength and high modulus), leading to greater strength in shear and compression and higher service-temperature capabilities. Thus, they have significant scientific, technological and commercial importance [5,6]. During the last decade, because of their improved properties, MMCs are being used extensively for high performance applications such as in aircraft engines and more recently in the automotive industry [7,8].MMCs offer a unique balance of physical and mechanical properties. Aluminium based MMCs have received increased attention in recent decades as engineering materials with most of them possess the advantages of high strength, hardness and wear resistance [9,10].

Fly ash particles are potential discontinuous dispersoids used in metal matrix composites, since they are low-cost and lowdensity reinforcement available in large quantities as a waste by-product in thermal power plants[10,11]. The fly ash contains the most important chemical constituents like SiO_2, Al_2O_3, Fe_2O_3 and CaO. It constitutes quartz, mullite, magnetite, hematite, spinel, ferrite and alumina[12]. Addition of fly ash particles in to Al matrix improves the hardness, wears resistance, damping properties, stiffness and reduces the density [13].The ductility of the composite decreases with increase in the weight fraction of reinforced fly ash and decreases with increase in particle size of the fly ash. However, for composites with more than 15% weight fraction of fly ash particles, the compressive strength is reported to be decreasing [14,15].

Most of the previous studies carried out on processing of aluminium–fly ash composites have utilized different size of reinforcement, different amount of reinforcement. But very little work has been reported on different amount of reinforcement in analysis of physical test. To overcome that situation, in this present investigation, we have casted two types of metal matrix composites by varying the wt. % of fly ash. As fly ash is cheap, abundant and it is a waste material from thermal power plant. The objective of the investigation was to compared the mechanical and tribological properties of the metal matrix composites with the base alloy.

2.EXPERIMENTALPROCEDURE

2.1 MATERIAL

Al–6061 alloy was used as a matrix. The composition of the alloy is given in Table 1. Cenospheres of fly ash were used as a reinforcement material in this investigation. They were formed in the temperature range of 920–1200 °C (32) .In this experiment we have used fly ash whose composition is as shown in the table-2.

Table-1: Composition of Al-6061 alloy [wt. %] designated asa base alloy

AL	Mg	Fe	Si	Cu
96.63	1.0	0.7	0.6	0.27
Zn	Cr	Ti	Mn	Others
0.25	0.2	0.15	0.15	0.05

Table-2: Chemical Composition of Fly Ash

Compound	**SiO_2**	**Al_2O_3**	**Fe_2O_3**	**CaO**	**MgO**	**Na_2O**	**K_2O**
(Wt. %)	63.34	24.60	4.97	1.23	0.56	0.11	0.64

2.2 PREPARATION OF ALUMINIUM 6061 ALLOY BASED METAL MATRIX COMPOSITE BY STIR CASTING

After cleaning Al 6061 ingot, it is cut to proper sizes, weighed in requisite quantities and were charged into a vertically aligned pit type bottom pored melting furnace (Fig.1).

Fig.1: Bottom Pouring Furnace

Fly ashes were preheated to 650°C±5°C before pouring into the melt of Aluminium-6061 Alloy. This was done to facilitate removal of any residual moisture as well as to improve wetability. The molten metal was stirred with a BN coated stainless steel rotor at speed of 400-450 rpm. A vortex was created in the melt because of stirring where preheated fly ash was poured centrally in to the vortex. The rotor was moved down slowly, from top to bottom by

IOP Conf. Series: Materials Science and Engineering **410** (2018) 012016 doi:10.1088/1757-899X/410/1/012016

maintaining a clearance of 12mm from the bottom. The rotor was then pushed back slowly to its initial position. The pouring temperature of the liquid was kept around 700^{0}C. Casting was made in rectangular metal mould (250x20x45 mm^3). For comparison purpose two composites were prepared with 10 and 15 wt.% of fly ash.

2.3 Mechanical Properties

2.3.1 IMPACT STRENGTH TEST

Impact Strength tests were performed by Charpy V Notchpendulum impact testing machine as shown in Fig 2. The square bar test specimens were placed as simply supported beams. Specimens were prepared by square cross section 10 mm x 10mm and 55 mm in length with 45-degree v notch at the centre as shown in Fig. 3. Single blow of hammer was given at mid span of specimThe sufficient blow was applied to bend or break the specimen at centre. The striking energy was measured as 310 ±10 joules.

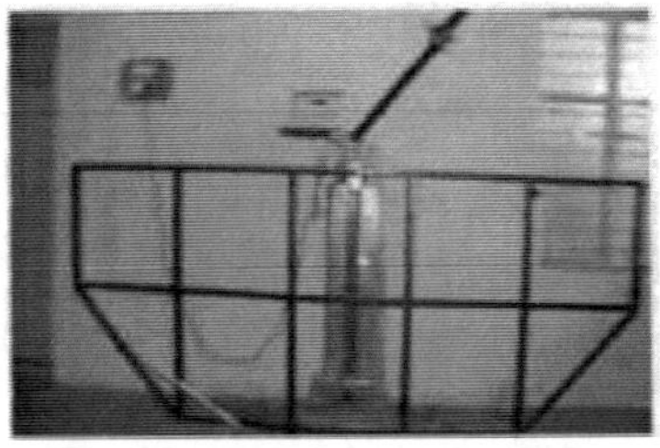
Fig.2 Impact testing machine

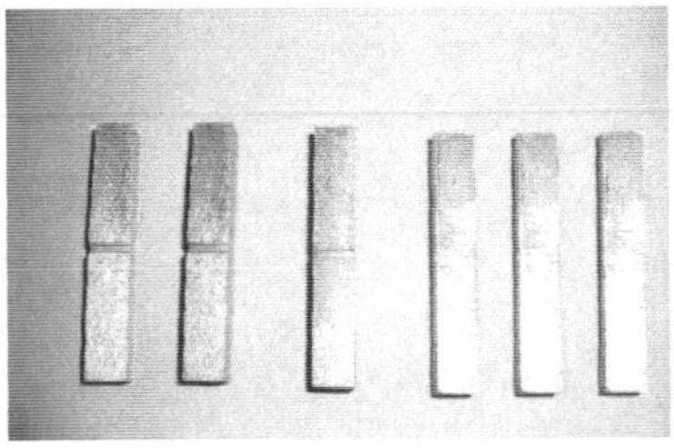
Fig.3 Specimen before impact test

Fig.4 Specimen after impact test

2.3.2 COMPRESSION STRENGTH TEST

Compression test was done in the Universal Testing Machine (UTM) shown in Fig5. The cylindrical test specimen was mounted on the base plate of the UTM. The specimen here used had same diameter as that of height of the specimen shown in Fig6. The load was applied on the specimen gradually until the sample was compressed until its height reduced by 50%. As the application of load increased the displacement also increased, then the displacement reduced drastically as it cannot be compressed more. The photograph of used UTM and test specimens is shown in Fig below.

Fig.5 UTM machine

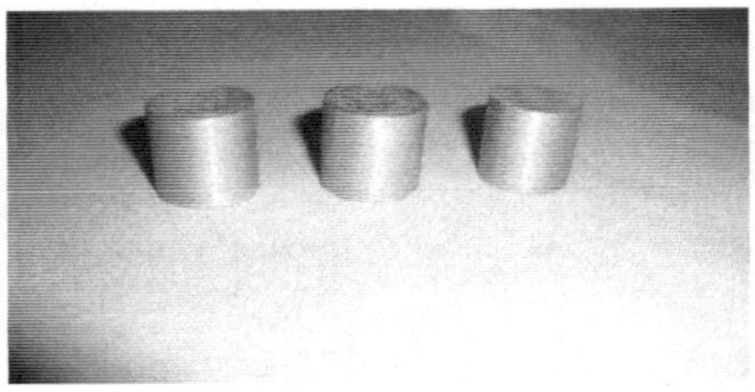
Fig.6 Specimen before compressive test

Fig.7 Specimen after compressive test

3. RESULT AND DISCUSSION

3.1 MECHANICAL PROPERTIES

Charpy Impact data for the three materials i.e. Al-6061 alloy, MMC with 10% fly ash and MMC with 15% fly ash were depicted in Table3. It was observed that the amount of energy absorbed by MMC prepared with 15% fly ash is higher in comparison to the other two materials i.e. Al-6061 alloy and MMC prepared with 10% fly ash. The

IOP Conf. Series: Materials Science and Engineering **410** (2018) 012016 doi:10.1088/1757-899X/410/1/012016

compressive test result was listed in Table 4. Data from compression test had also shown the highest strength value for MMC prepared with 15%fly than the other two materials. From the table-4, it can be observed that the compressive strength increased with an increase in the weight percentage of fly ash particles. This is due to the hardening of the base alloy by fly ash particles

Table-3: Impact Strength of Al-6061, MMC with 10% fly ash and MMC with 15% fly ash.

Sample no.	Sample name	Trial no.1	Trial no.2	Trial no.3	mean
01.	Al-6061	0.71	0.72	0.70	0.71
02.	MMC with 10% fly ash.	1.63	1.63	1.62	1.62
03.	MMC with 15% fly ash.	1.81	1.82	1.81	1.81

Table-4: Compressive Strength of Al-6061,AMC with 10% fly ash and AMC with 15% fly ash.

Sample No.	Sample Name	Trial no.1	Trial no.2	Trial no.3	Mean Compressive Strength In N/mm^2
01.	Base alloy	460	462	459	460
02.	MMC with 10% fly ash.	478	479	479	479
03.	MMC with 15% fly ash	483	482	484	483

4.CONCLUSION

- Stir casting method was used to prepare the composite which result in uniform distribution of flyash particles.
- Al-6061 alloy matrix composite was successfully fabricated with uniform distribution of flyash particales.
- Impact strength of the material was found to be increased with the increase in flyash concentration.
- It was observed that due to addition of flyash particles Compressive strength of the composite was increased.
- Physical properties of the Al-6061 alloy matrix was inhanced due to the addition of flyash which ultimately inhanced due to addition of flyash which ultimately inhanced mechanical properties of the metal matrix composite.

IOP Conf. Series: Materials Science and Engineering **410** (2018) 012016 doi:10.1088/1757-899X/410/1/012016

5.REFERENCES

[1] Feest, E. A. *Materials & Design;* 1986, 7, 2, 58-64.

[2] Miracle, D.B. *Composites Science and Technology,* 2005, 65, 2526-2540.

[3] M.F. Ashby and D.R.H, Jones, Elsevier Science and Technology, 1980.

[4] Ashby, M. F. and Jones, D. R. H. UK: Elsevier Science and Technology, 1980.

[5] Pedersen, W and M.Ramulu Journal of Materials Processing Technology, 2006,172, 417–423.

[6] Day, R. D.; R.F. Smart and N.Tommis 1970, U.K., Paper No.4.

[7] Suresh, S.; A.Mortensen and A. Needleman Butterworth-Heinemann, Stoneham; MA02180, USA, 1993.

[8] Yan, C.andL.C.Zhang; Applied Composite Materials, 1995, 1, 431–447

[9] F. Toptan, A. Kilicarslan , and I. Kertil, Materials Science Forum, 636-637, pp.192-197, 2010.

[10] Pramanik, A.; L.C.Zhang, and J.A.Arsecularatne *International Journal of Machine Tools and Manufacture,* 2006, 46, 1795–1803.

[11] C. Limmaneevichitr, W. Eidhed, Materials Science and Engineering, vol. 355A, pp.174–179, 2003

[12] F. Toptan, A. Kilicarslan , and I. Kertil, Materials Science Forum, 636-637, pp.192-197, 2010.

[13] Pramanik, A.; L.C.Zhang, and J.A.Arsecularatne, *International Journal of Machine Tools and Manufacture,* 2006, 46, 1795–1803a.

[14] Z. Zhang,J. Zhang, and Y.W. Mai, vol.176, pp. 231-7, 1994.

[15] C. Limmaneevichitr, W. Eidhed , Materials Science and Engineering, vol. 355A, pp.174–179, 2003.

IOP Conf. Series: Materials Science and Engineering **410** (2018) 012017 doi:10.1088/1757-899X/410/1/012017

Study and characterization of bio-active glass coating composite with and without hydroxyapatite on titanium and SS316L to regenerate supporting bony growth to established better bonding and stability

Tarak Das[1], Pratik Das[2] and Piyali Basak[2]

[1] *Dept, of Biomedical Engineering, Netaji Subhash Engineering College, Garia, Kolkata*

[2] *School of Bioscience and Engineering, Jadavpur University, Kolkata*

ABSTRACT

Over a period of time a number of biomaterials have been developed and are used for hard tissue and joint. Metal Biomaterials gained its importance in recent time for having advantageous mechanical properties, nontoxic behavior, and biocompatibility. But one of the main drawbacks of using metallic implants is that they are bio-inert and inside the body they become encapsulated by dense fibrous tissue. This impedes proper stress distribution at the implant bone interface, which can result in an interfacial failure and loosening of the implant and finally lead to crack or fracture in the adjacent bone. Bone cement (Poly Methyl Methacrylate) can be used to improve the implant fixation but it becomes brittle with time so it gets loosen from the surface as well as it has a poor solvent resistance. Calcium Phosphates, specially Hydroxy-Apatite [$Ca_{10}(Po_4)_6(OH)_2$] have been widely used as coating materials on metallic implants due to the bio-active nature. Plasma-spray technique, in particular, has been widely investigated by many researchers and has been proposed as a reasonable coating method. But during the plasma spraying process, high flame temperature is required which can lead to compositional changes and structural transformation of Hydroxy Apatite (HA). So, after this procedure the uniformity of physical and chemical properties as well as bio-activity of Hydroxy Apatite (HA) could be lost. In this study one bio glass composition was prepared by matching its thermal expansion coefficient nearer to thermal expansion coefficient of metallic substrate specially Titanium and SS-316L. The metallic substrate coated with this Bio glass with and without mixing with HA with a certain percentage. The coating adherence property was studied by 3-point bending moment and bioactivity studied by using Simulated Body fluid which showed very promising result. The coating was also characterized with analysis of DTA-TGA, Surface reflectance also Surface Roughness.

Keywords: Hydroxy Apatite, SS-316L, Glass Coating, 3 point Bending-Moment, DTA-TGA, Surface Reflectance, Surface Roughness.

*Corroseponding E-mail: tarakdas111@gmail.com

IOP Conf. Series: Materials Science and Engineering **410** (2018) 012017 doi:10.1088/1757-899X/410/1/012017

1. INTRODUCTION

During last few decade there have been developed of several bioactive coating materials[1] by different techniques for using orthopedical implants to increase better bonding and stability. Calcium Phosphates, specially Hydroxy apatite [$Ca_{10}(PO_4)_6(OH)_2$] have been widely used as coating materials on metallic implants due to its bio-active nature to facilitate bonding with bone.

Plasma-spray technique, in particular, has been widely investigated by many researchers and has been proposed as a reasonable coating method. But during the plasma spraying process, high flame temperature is required which may lead to compositional changes and structural transformation of Hydroxy apatite (HA).So, after this procedure the uniformity of physical and chemical properties as well as bio-activity of Hydroxy apatite (HA) could be lost. Most implants today are made from either titanium-based alloys or alloys made from a mix of cobalt and chromium. Both possess excellent mechanical properties but neither is able to bond with bone. As a result, these metals rub against the bones into which they've been implanted, creating wear and tear that shortens implant lifetimes.In many research work bone cement (Poly methyl methacrylate) have been used for better bonding and stability but it may result in deterioration of the adjacent bone and failure.

The aim of the present work is to develop Bio-active glass coating which is to be used for orthopedic dental implants to improve the quality of life for people who have lost their teeth. At present, in India as well, Central Glass and Ceramic Research Institute (C.G.C.R.I), Kolkata, is involved in development of titanium base dental implants which is to be coated by Bio-active glass With and without Hydroxy Apatite (HA)

2. EXPERIMENTAL PROCEDURES

The composition of the coating materials was determined in such a way so that it's calculated thermal expansion coefficient (Using Winkelmann and Schott factors) can match with the metal substrate[3]. The glass was prepared using a conventional glass melting procedure. The appropriate amounts of reagents/raw materials quartz (SiO_2), $CaCO_3$,[2] dry soda ash (Na_2CO_3), decahydrated borax ($Na_2B_4O_7$,$10H_2O$), TiO_2, di-amonium hydrogen phosphate[$(NH_4)_2HPO_4$], were mixed together properly in water. The batch composition of the glass was shown in table-IV.3. The mixture was first dried at 100°C for 1 hr then melted in air at temperature 1600°C for 3 hrs. in a Pt crucible. The experimental variables used during glass melting were shown in table-IV.4. The melt was quenched in to water and obtained flaky glass particles, dried and stored for subsequent use. The frit was subsequently milled along with different mill additives in aqueous medium in a porcelain jar for 72 hrs with HA (20%) and without HA. The oxide composition [4] and batch composition of prepared glass given in table: 1 and table: 2.

Table: 1. Oxide composition of the prepared glass (mol %)

SiO_2	Na_2O	CaO	P_2O_5	B_2O_3	TiO_2
59 - 61	8 - 12	21 - 25	2 - 5	2 – 5	0.5 – 2.5

Table: 2. Batch composition of the prepared glass (wt %)

SiO_2	$Na_2B_4O_7$, $10H_2O$	Na_2CO_3	$CaCO_3$	$(NH_4)_2HPO_4$	TiO_2
42 - 44	5 - 7	10 - 13	28 - 331	7 – 10	0.5 – 2.5

Titanium [5] metal and SS316L was coated by the HA free glass and glass with 20% HA [6]. The coated sample is depicted in Fig: 1, Fig: 2, Fig: 3 and Fig: 4.

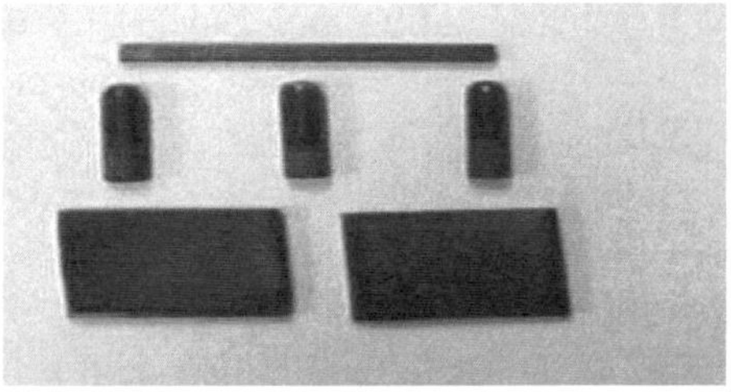

Fig: 1. Glass Coated (without HA) Ti Samples

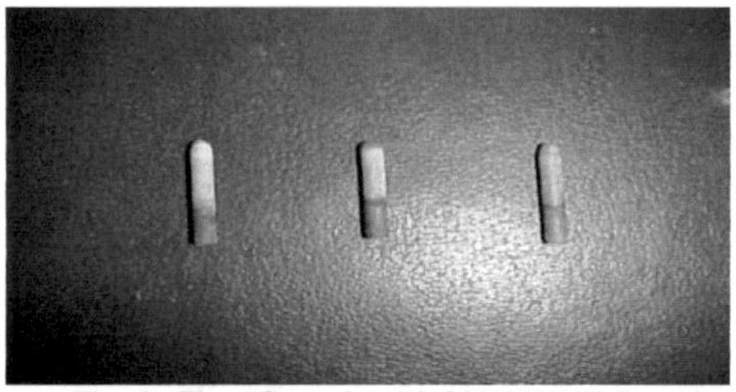

Fig: 2. Glass Coated (with 20%HA) Ti Samples

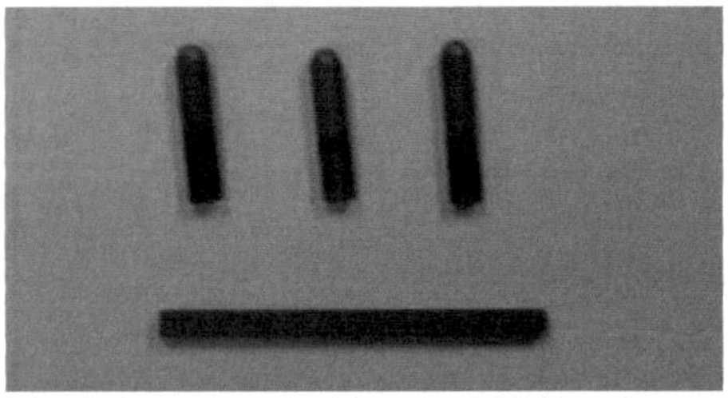

Fig: 3. Glass Coated (without HA) SS-316L Samples

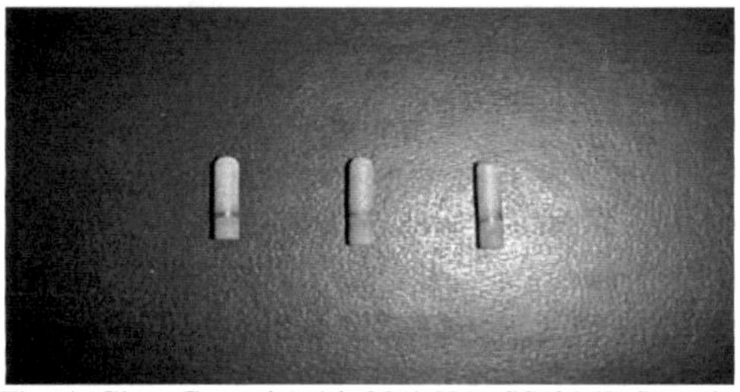

Fig: 4. Glass Coated (with 20% HA) SS-316L Samples

3. Results and discussion

3.1: DTA/TGA ANALYSIS

The Fig: 5 show a typical DTA/TGA pattern of the bioactive glass composition [7] investigated. The TGA graph indicates that the glass is thermally stable and no significant wt. loss is observed during heating to 800°C.

The DTA pattern showed one endothermic peak at 41.0°C probably due to moisture removal, one exothermic peak at 416.6°C, which may be due to structural changes, and one endothermic nucleation temperature at 631.8°C and exothermic growth temperature at 756.5°C.

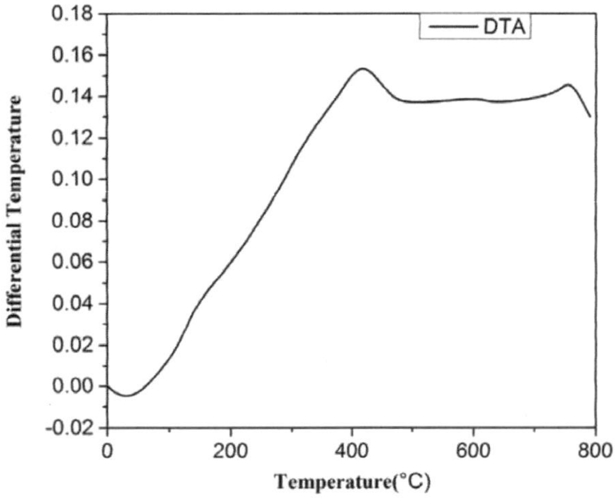

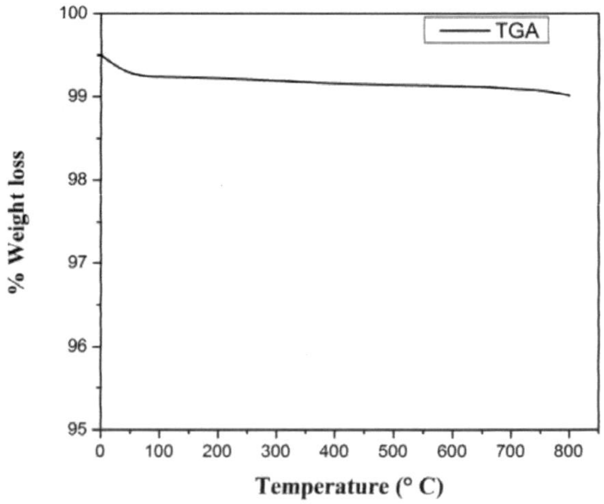

Fig: 5. DTA/TGA Analysis of bioactive glass (without HA) coated sample

3.2: SURFACE ROUGHNESS:

The surface roughness (Ra in micron) of coated sample (with & without HA) is depicted in table-3. Qualitatively we analyzed that, the coating with low Ra value having better adherence to the metal surface. It may be due to proper fusion and settling of molten glass coating material to the metal surface.

Table: 3. Surface Roughness value of coated sample

Name of the Sample	Average Ra Value(in micron)	Standard Deviation(in micron)
SS-316L without HA	2.406	0.397
SS-316L with 20% HA	1.569	0.284
Ti without HA	1.797	0.338
Ti with 20% HA	1.806	0.370

3.3: SURFACE REFLECTANCE:

Surface reflectance of coated samples were measured using Double Beam Ratio Recording Spectrophotometer Model no.UV3101PC.Results of surface reflectance measurements of the coated samples indicate that the average surface reflectance is in the range of 0.1-0.7% in the range of 500-1500 nm wavelength of radiation. The results are in agreement with the surface roughness measurement data.

3.4 RESULTS OF IN-VITRO ANALYSIS OF COATED SAMPLE IN SBF SOLUTION

The change in weight of the coated samples with time of duration of immersion in SBF [8] solution is shown in table 4

Table: 4. Change in weight of coated Ti and SS-316L samples in Simulated Body Fluid

Sample name	Weight(g)				
	Initial	After 1 week	After 2 weeks	After 3 weeks	After 4 weeks
TB	2.2079	2.2079	2.1978	2.1970	2.1967
TH	2.2271	2.2271	2.2271	2.2270	2.2170
TU	2.1754	2.1755	2.1756	2.1757	2.1859
SB	2.5462	2.5462	2.5459	2.5456	2.5353
SH	2.2392	2.2392	2.2389	2.2385	2.2382
SU	2.3384	2.3385	2.3385	2.3485	2.3487

TB & SB: Titanium & Stainless steel coated with Developed Bioactive glass
TH & SH: Titanium & Stainless steel coated with Developed Bioactive glass containing 20% HA
TU & SU: Uncoated Titanium & Stainless steel

The weight change data (Fig: 6) indicates that the reaction of the coating with the S.B.F fluid was initially very negligible but with progressive duration of exposure the reaction rate increases resulting in increasing weight loss. The initial inertness of the coating may be attributed to the presence of a silica network on the as fired coating surface. Once the reaction was initiated the reaction rate becomes almost constant throughout the rest of the period as was reflected in the change in weight values. The results also indicate that incorporation of HA lower down the reaction rate of the resultant coating. The metal controls used are naturally very less reactive.

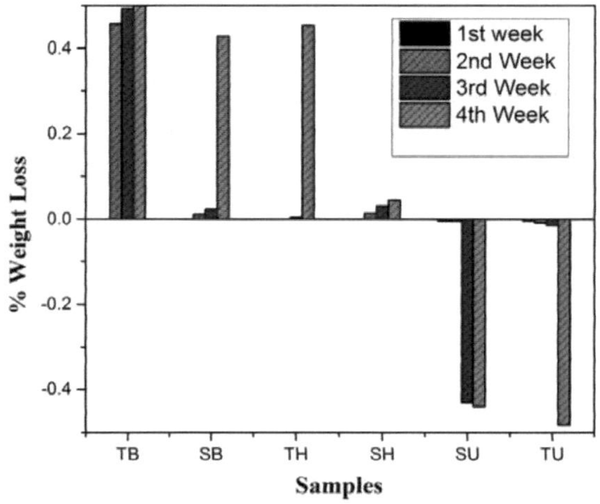

Fig: 6 Change in Weight loss

3.5 COATING ADHERENCE:

The adherence of Developed Bioactive glass (with or without HA) coating on Ti and SS-316L metal surface [9] was tested by 3-Point Bending Moment using INSTRON 5500R [10]. The result is shown in table: 5. the results indicate that the coatings are strongly adhered to the metal substrates.

Table: 5. Results of thee point bending moment of coated samples

Name of the samples	Load to break the coating (N)	Stress to break the coating(MPa)
Bioactive Glass Coated Ti	460 – 580	700 – 860
Bioactive Glass Coated SS-316L	460 – 580	680 – 720
HA containing Bioactive Glass Coated Ti	430 – 540	575 – 750
HA containing Bioactive Glass Coated SS-316L	420 – 530	525 – 630

4. CONCLUSIONS

The work is still under progress but till its output showing the attractive results to reach our target for developing better strengthening bioactive glass coating. The in vitro behavior of the silicate glass coatings analyzed in this work is similar to that of bulk glasses. Coatings with silica content lower than 60wt% precipitated apatite during in vitro testing. The mechanism of apatite formation is similar to that described by Hench for Bioglass®. However, due to their lower silica content, the thermal expansion of these coatings is higher than that of Ti and the tensile thermal stresses generated during processing drove the slow growth of cracks in S.B.F. The cracks eventually reached the interface and initiated delamination. Higher silica coatings did not form apatite but were more resistant to corrosion and slow crack growth. At the moment, graded coatings that have glasses with high silica content in contact with the metal and a low silica glass on their surface are under

IOP Conf. Series: Materials Science and Engineering **410** (2018) 012017 doi:10.1088/1757-899X/410/1/012017

development in order to improve their long term stability while maintaining a good biological response.

HA particles were incorporated into these coatings to promote increased bioactivity. The effectiveness of Bio-glass incorporation depended on the softening temperature of the glass coating, with higher softening temperatures leading to increased degradation of the Bio-glass. This was not the case for HA. Incorporation of ~20% of either particle type was successfully accomplished without cracking or loss of adhesion to the melted substrate

5. REFERENCES

[1] L.L. Hench and O. Andersoon, 1993, Bioactive glass coatings In: L.L. Hench and J. Wilson Editors An introduction to bioceramics World Scientific New Jersey pp 239-260.

[2] A. Pazo, E. Saiz and A.P. Tomsia 1998, Acta Materials, Vol. 46, pp 2551-2558.

[3] L.L. Hench, 1991, Journal of American Ceramic Society, Vol. 74, pp 1487-1570.

[4] M. Brink, T. Turunen, R.P. Happonen and A. Yli-Urpo 37, 1997, Journal of Biomedical Materials Research, Vol. 12, pp 114-121.

[5] J.M. Gomez-Vega, E. Saiz and A.P. Tomsia, 1999, Journal of Biomedical Materials Research, Vol. 46, pp. 549-559.

[6] T.das, N.Banerjee and P.Basak 2017, International journal of engineering technology management and applied sciences pp 82-95.

[7] L.L. Hench Bioceramics: from concept to clinic, 1991, J Journal of American Ceramic Society, Vol. 74, pp 1487-1570.

[8] P.N. de Aza F. Guitian A. Merlos E. Loratamayo and S. de Aza, 1996, Journal of Materials Science, Vol. 7, pp. 399-402.

[9] V.V.Vargin and Kenneth Shaw, Technology of Enamels, pp. 38-49

[10] T. Das N.Banerjee and P. Basakl 2017 Development and analysis of glass & ceramic based coating material for orthopaedic implants to increase better bonding and stability International Journal of innovative research in science engineering and technology An ISO 3297: 2007 Certified Organization pp: 1042-1052.

IOP Conf. Series: Materials Science and Engineering **410** (2018) 012018 doi:10.1088/1757-899X/410/1/012018

Phytochemical analysis and comparative study of antibacterial effect of turmeric extracts using different solvent.

Dr. Piyali Basak, Tathagata Adhikary, Pratik Das, Shreya Biswas
School of Bioscience and Engineering, Jadavpur University, Kolkata, West Bengal 700032

Abstract:
Curcumin, the principal curcuminoid found in turmeric, is generally considered its most active constituent. Curcumin, besides its anti-inflammatory property, has been known to possess *in vitro* anti-microbial potential against a wide range of microorganisms including fungi as well as several Gram-positive and Gram-negative bacteria. Curcumin possesses a synergistic effect with important antibiotics also. The mechanism of antibacterial activity of curcumin seems to differ depending on the strain being studied. Coating of conventional wound dressing materials with curcumin composite can enhance the effectiveness of the material. Studies have shown that fabrication of silver nano-composite films impregnated with curcumin showed the stronger antibacterial activity against E. coli. This research would aim to claim the best solvent used in extraction process of curcumin and demonstrate the antimicrobial effect of curcumin in order to develop a curcumin-nano-composite coated material for wound dressing with increased efficiency.

1. Introduction

Turmeric belongs to the ginger family, Zingiberaceae. It has been traditionally used as a spice and medicine from ancient times and its potential in other fields has been harnessed in recent years. Phytochemical components of turmeric include sugars, proteins, resins, traces of volatile oils and a compound called curcuminoids which includes curcumin (diferuloylmethane), demethoxycurcumin, and bisdemethoxycurcumin. Major reported pharmaceutical activites of turmeric are antioxidant;antibacterial, anti-inflammatory, anti-tumor and anti-cancer activities and these are due to the presence of curcumin in it, hence claiming it to be the most bioactive component in turmeric. Curcumin is generally regarded as safe (GRAS) by the United States Food and Drug Administration (FDA) due to its low toxicity even when ingested at relatively high levels. Curcumin is a phenolic constituent and is hydrophobic in nature [1]. So it can be extracted efficiently by many organic solvents. Curcumin, on an average, occupies 3.14% w/w of powdered turmeric, and this percentage varies with the species of *Curcuma longa.* E.coli and S. aureus are the two bacterias mostly found in different types of wounds and play a role in delaying the wound healing process resulting wound infections [2]. In a study, the predominant bacteria isolated from the infected wounds were Staphylococcus aureus 47 (32.4%) followed by Escherichia coli 29 (20%), Proteus species 23 (16%), Coagulase negative Staphylococci 21 (14.5%), Klebsiellapneumoniae 14 (10%) and Pseudomonas aeruginosa 11 (8%) [3]. Turmeric extracts showed a prominent zone of inhibition against E. coli and S.aureusmaking it an excellent wound healing agent due to its non-toxic nature. The synergistic activity of curcumin with other antibiotics like ampicillin oxacillin, and norfloxacinhas been exploited in recent years which showed a remarkable decrease in the minimum inhibitory concentration of the antibiotics against bacterial strains [4].

So we have extracted phytochemicals from turmeric using different solvents and then compared antibacterial effect of different extracts.

2. Materials, reagents and solvents used:

Turmeric, foiln-ciocalteau reagent (FCR), gallic acid, quercetin, nutrient broth, agar agar, streptomycin, benzene, acetone, acetonitrile, ethanol, 2-propanol, methanol, dimethyl sulfoxide, silver nitrate.Test Microorganisms: Standard strains of E.coli and S. aureus

3. Methods performed:

3.1 Preprocessing of the turmeric: Before the extraction process, raw turmeric was washed with water, peeled, cut into small pieces and was dividedinto two parts- one sundried for two weeks (part A) and the other was kept in the incubator at 37 °C for two days (part B) to reduce the moisture content in the turmeric to a negligible amount. Then it was crushed into coarse powder form using mortar and pestle.

3.2. Extraction of the bioactive components: 30 grams of powdered turmeric from part A and part B was macerated with 200 ml of methanol for 3 days allowing enough time for the bioactive components to solubilize in the solvent. This was followed by digestion. Digestion was performed by adding fresh 10 grams of powder in the solution with gentle heating to increase the concentration of the bioactive components in the solution and is left for another one day. Similar process is followed for the turmeric powder from part B with solvents namely benzene, acetone, 2-propanol, ethanol, and acetonitrile.

3.3.Sample preparation: The crude extracts were filtered using Whatman No.1filter paper and concentrated using rotary evaporator. Then the highly concentrated solutions were lyophilized to achieve the final prepared samples which were stored at -20 °C for further use. Samples (i.e. the final product after lyophilization) were named from A to F depending on the solvent used in the extraction process where A, B, C, D, E, F corresponds to extraction by benzene, acetone, 2-propanol, ethanol, acetonitrile and methanol respectively.

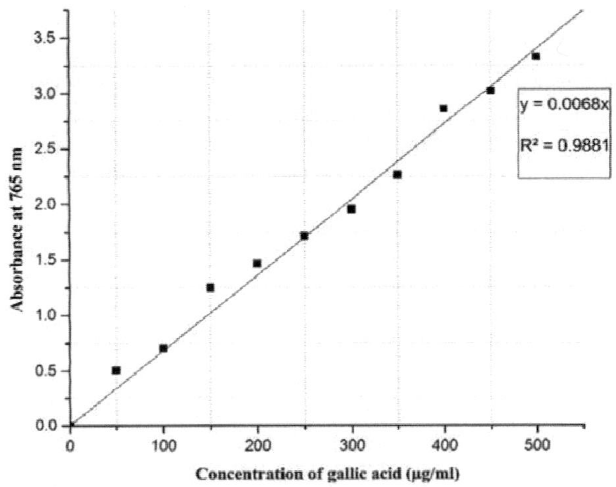

Figure 1: Linear curve of Absorbance vs Concentration of standard graph of Gallic acid

3.4. Phytochemical analysis:

3.4.1 Total phenolic content (TPC): TPC of each lyophilized samples were calculated from the calibrated standard curve using gallic acid by spectrophotometric analysis using FolinCiocalteu reagent (FCR). In this process, 0.5 ml of extracts (of concentration 1000 µg/ml) or gallic acid (of different concentrations from 0-500 µg/ml) were mixed with 3 ml of distilled water. Then 0.25 ml of FCR was added followed by 0.75 ml of saturated aqueous sodium carbonate solution. Lastly, 1 ml of distilled water was added and the mixture was incubated for 30 minutes in 37 °C. Absorbance is taken at 765 nm against a selected blank using UV-visible Spectrophotometer [5].

A standard curve of gallic acid is prepared at-first as reflected in Figure 1. Phenol content of different samples was calculated in gallic acid equivalent.

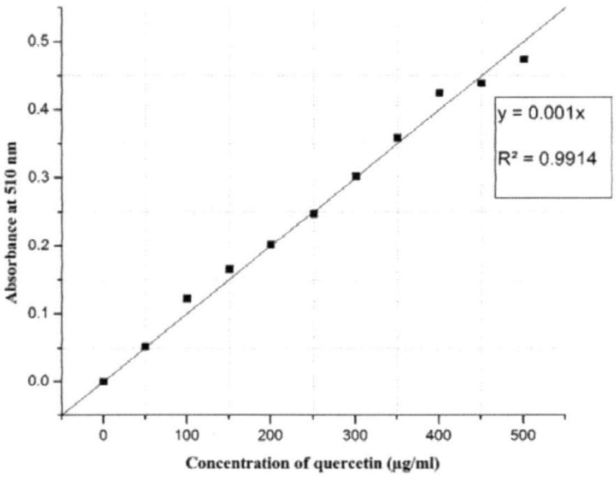

Figure 2: Linear curve of Absorbance vs Concentration of standard graph of Quercetin

3.4.2 Total flavonoid content (TFC): TFC of each extracted samples are calculated from the calibrated standard curve using quercetin by spectrophotometric analysis. In this process, 0.5 ml of extracts (of concentration1000 µg/ml) or quercetin (of different concentrations from 0-500 µg/ml) were mixed with 0.5 ml of distilled water. Then 0.3 ml of 5% sodium nitrite was added and kept at room temperature for 5 minutes after which 0.3 ml of 10 % aqueous solution of aluminum chloride was added. 5 minutes later 2 ml 1 molar sodium hydroxide solution was added and test tubes were shaken before taking the absorbance at 510 nmagainst a selected blank using UV-visible Spectrophotometer [6].

A standard curve of quercetinis prepared at-first as reflected in Figure 2. Flavonoid content of different samples was calculated in quercetinequivalent.

*3.5. Nanocomposite synthesis:*As sample F has highest phenol and flavonoid content we choose it for further study. 100 mg of sample F was dissolved in 20 ml of dimethyl sulfoxide and added to an aqueous solution of silver nitrate (0.1 % w/v) dropwise under constant stirring condition. Stirring was continued for 6 hours resulting in the formation of a clear brown colored solution. Solution was centrifuged at 6000 rpm for 20 minutes. Supernatant was discarded and the pellet so formed was washed with distilled water before collecting it in a petri plate for freeze drying. The nanocomposite(sample G) synthesized using the turmeric extract was characterized by FTIR and FESEM [7].

3.6.Antibacterialassay: To check the potency of turmeric extracts against selected bacterial strains, 50 mg of each of the lyophilized samples and 20 mg of the synthesized nanocompositewere dissolved in 1 ml of DMSO, DMSO being the standard solvent for each samples. Both well-diffusion and disc-diffusion methodswere used to determine the antibacterial activity against E.coli and S. aureus. DMSOwas used as the control and streptomycin as the standard. Plates are incubated for 24 hours and the zone of inhibition was measured. Then, the mean value of two different methods was taken into account. There was not much significant difference between two values.

4. Results and discussions:

On the first trial of extraction with methanol, both the samples designated as part A and part B showed almost same results of phytochemical analysis. So turmeric powder from part B was chosenhenceforth for the extraction process by different solvents due to its shorter preparation time.

Table 1: Total phenol content expressed in gallic acid equivalent (mg GAE /g of extract)

Sample	Absorbance at 765 nm	TPC (mg GAE / g)
A	0.846	248.79
B	0.913	268.64
C	0.929	273.12
D	1.085	319.18
E	1.132	332.91
F	1.356	398.68

Table 2: Total flavonoid content expressed in quercetin equivalent (mg QE / g of extract)

Sample	Absorbance at 510 nm	TFC (mg QE / g)
A	0.082	164.00
B	0.151	302.00
C	0.087	174.00
D	0.105	210.00
E	0.062	124.00
F	0.148	296.00

4.1. Total phenol content:

From Table 1 it is observed that methanolic extract showed the highest phenol content while benzene extract has the lowest value among others. Total phenol content varied significantly with the solvent used during the extraction process.

4.2. Total flavonoid content:

From Table 2 it is observed that Acetone extract has the highest flavonoid content followed by methanol whereas acetonitrile extract witnessed the least value. Again, the solvent used in the extraction greatly influenced the flavonoid content.

4.3. FTIR analysis of turmeric extract and its nanocomposite:

FTIR result of turmeric extract (sample F) and its comparison with the synthesized nanocomposite predicts the presence of curcumin in both the samples. The peak at 3338.78 $cm^{-1}$of turmeric extract corresponds to the alcohol/phenol O-H bond stretching and it gets shifted to 3523.95 in nanocomposite. Similar instances have been found where the peak values (in cm^{-1}) 2929.87, 1412.54, 1276.88, 1207.44, 1101.35, and 987.55 gets shifted to 2926.01, 1436.97, 1278.81, 1128.36 and 952.84 respectively. These peaks correspond to C-H stretch, aromatic C=C, enol C=O, phenol C=O, enol C-O-C and benzoate trans C-H vibration respectively. Shifting indicates the formation of co-ordination bond between silver atoms and electron rich groups present in the extract and thus claims a probability of their involvement in nanocomposite synthesis.

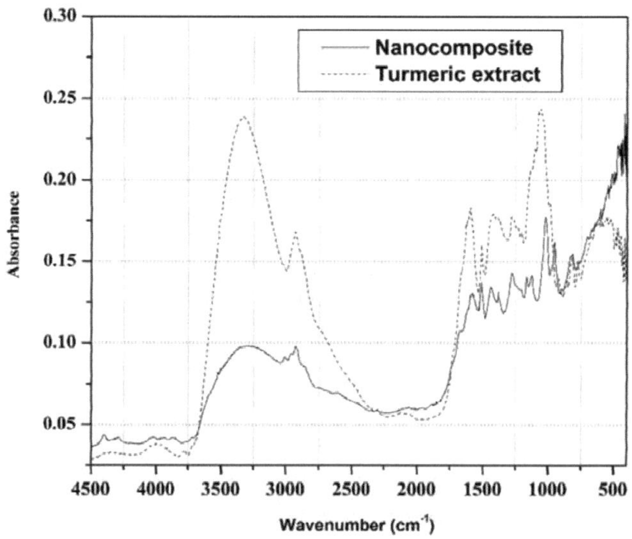

Figure 3: FTIR image of turmeric extract and its nanocomposite

IOP Conf. Series: Materials Science and Engineering **410** (2018) 012018 doi:10.1088/1757-899X/410/1/012018

\

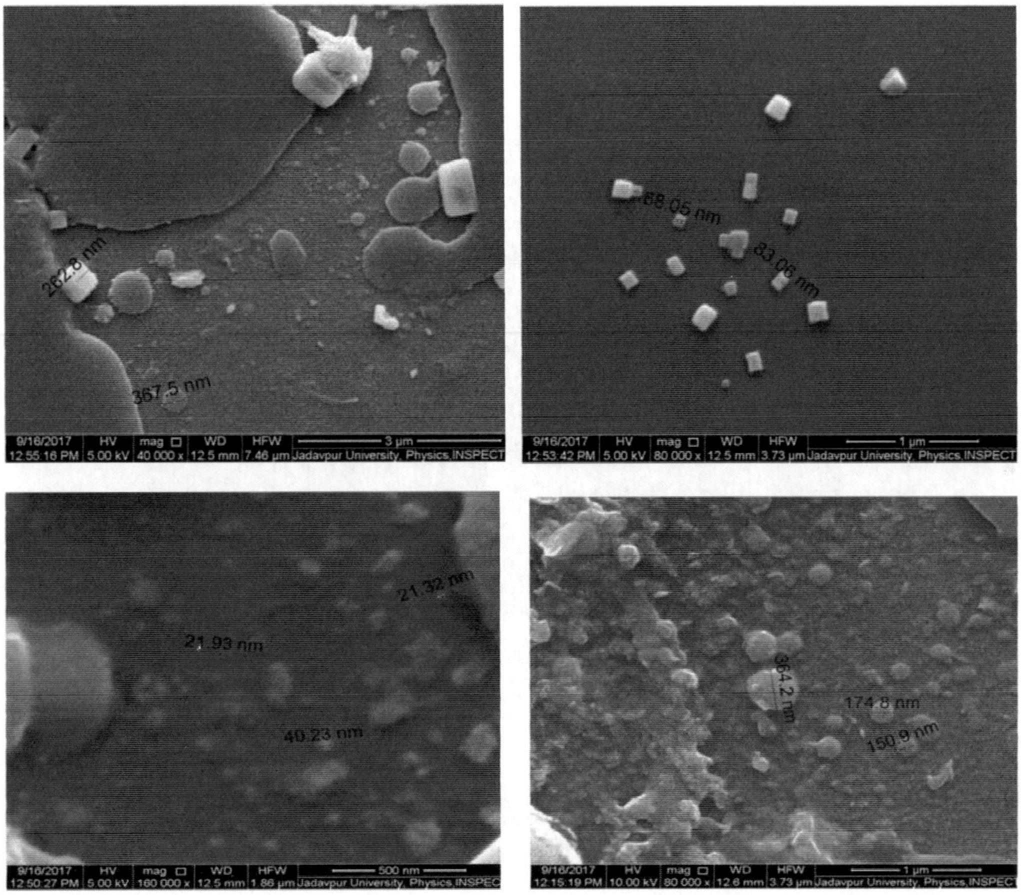

Figure 4: FESEM images of the nanocomposites

4.4. FESEM analysis of the nanocomposite:

The morphology of the nanocomposite was studied by FESEM as reflected in Figure 4. Two types of structure were found- spherical and cubical. Figure 4 shows that the nanoparticles were well dispersed and the particle size varied from 20 nm to 400 nm. Agglomeration is not visible.

4.5. Antibacterial activity:

Methanolic extract showed the highest antibacterial activity while benzene extract had the least potency against the selected strains of bacteria. It was observed that the total content of phenols plus flavonoids directly influenced the % inhibition, i.e. more the value of summation of TPC and TFC, more was the zone of inhibition. The synthesized nanocomposite showed high antibacterial activity even at much lower concentration than the extracts, thus claiming to enhance the activity of the bioactive components present in the extracts.

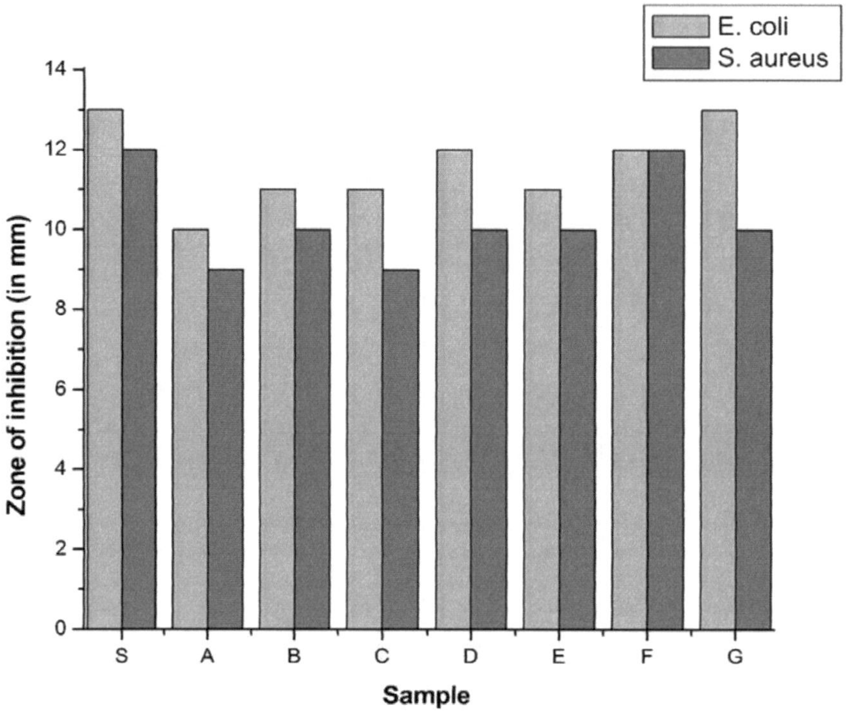

Figure 5: Zone of inhibition of different samples against different strains

5. Conclusion:

Six solvents were used in the extraction process of turmeric. TPC and TFC varied with the solvent used in extraction. The highest and the lowest TPC were observed in methanolic and benzene extract respectively. Acetone extract achieved the highest TFC while acetonitrile extract had the lowest value. Nanocomposite synthesized using turmeric extract and $AgNO_3$ was found to be existed in 2 shapes-cubical and spherical with sizes ranging from 20-400 nm. FTIR analysis confirms the interactions between $AgNO_3$ and turmeric extract in the nanocomposite. The antibacterial activity of different extracts seemed to be directly dependent on the summation of TPC and TFC. Nanocomposite showed a high potency against bacterial strains of E. coli and S. aureus at a much lower concentration than the extracts, thus enhancing the activity of the extracts.

6. References

[1] M.K. Nelson,J.L. Dahlin, J. Bisson, J. Graham, G.F. Pauli, and M.A. Walters, 2017, Journal of Medicinal Chemistry, Vol. 60, pp. 1620.
[2] P.G. Bowler, B.I. Duerden, and D.G. Armstrong, 2001, Clinical Microbiological Review, Vol.14, pp.244.
[3] M. Mama, A. Abdissa, and T. Sewunet, 2014, Ann Clinical Microbiology and Antimicrobials, Vol. 13, pp.1.
[4] S.Z. Moghadamtousi, H.A. Kadir, P. Hassandarvish, H. Tajik, S. Abubakar, and K. Zandi, 2014, BioMedical Research International, Vol.2014, pp.1.

[5] H. Mastura, Y. Hasnah, and T.N. Dang, 2017, International Food Research Journal, Vol. 24, pp/ 510.
[6] G. Li, S. Yu, Y.H. Zhou, and Q.F. Chen, 2013, Asian Journal of Chemistry, Vol. 25, pp.75.
[7] G.D. Venkatasubbu, T. Anusuya, 2017, International Journal of Biological Macromolecule, Vol. 98, pp. 66.

Green Synthesis and Characterization of Gelatin-PVA Silver Nanocomposite Films for Improved Antimicrobial Activity

Piyali Basak[1], Pratik Das[1], Sheya Biswas[1], Narayan Chandra Biswas[2], Gour Krishna Das Mahapatra[2].

[1]School of Bio-Science and Engineering, Jadavpur University, Kolkata -700032, India

[2]Department of Basic Engineering Science Netaji Subhas Engineering College, Kolkata`-700152

ABSTRACT

The present study deals with the in-situ production of gelatin-poly(vinyl alcohol)-silver nanocomposite films in view oftheir growing applications as antimicrobial packaging/container, wound dressing and antibacterial materials. Silver in the form of silver nanoparticles has made an amazing comeback as a prospective antimicrobial agent. The use of silver nanoparticles is also significant, as several pathogenic bacteria have developed resistance against various antibiotics. A unique, nontoxic, simple, lucrative and ecofriendly technique was used to synthesize green silver nanoparticles (AgNPs). The AgNPs were synthesized using Lupulus amarus extract as a reducing agent for silver nitrate salt (AgNO3). The particle size distribution of AgNPs was examined by Dynamic Light Scattering (DLS) and the concentration was examined by UV-VIS Spectrophotometer. The stable dispersion of silver nanoparticles was added slowly in gelatin –PVA solution and was crosslinked using glutaraldehyde (cross-linker). The Gelatin-PVA Silver nanocomposite solution was casted in a petri dish and dried to form a film. The nanoparticles encapsulated within polymer chains were characterized by X-ray diffraction (XRD) and Scanning Electron Microscopy (SEM). The green AgNPs nanocomposite film exhibited significant antimicrobial activity against both Gram-negative bacteria, and Gram-positive bacteria. Therefore, the present study clearly provides an approach to develop novel antimicrobial films which are possibly useful in preventing/treating infections and can be used as antibacterial container or in food packaging.

Key words: Nanocomposites, Crosslinking, Polyvinyl alcohol, Nanoparticles

Corresponding E-mail –pratik.me.online@gmail.com

1. INTRODUCTION

The nanomaterials can be produced by diverse approaches which include chemical, physical, and biological methods. The advance of new chemical or physical methods has resulted in environmental adulteration since the chemical trials involved in the production of nanomaterials produce a large quantity of dangerous and hazardous byproducts [1]. Thus, there is a necessity for "green nanotechnology" that embraces an uncontaminated, safe, environment-friendly, and non-toxic method of nanoparticle production, without the use of high pressure, energy, temperature, and toxic chemicals [2]. The biological approaches comprise production of nanomaterials from the extracts of plant, bacterial, fungal species [3].

Mankind is often infected by micro-organisms such as bacteria, virus, yeast, mold, etc [4]. Silver and silver ion based resources are extensively recognized for their bactericidal and fungicidal activity. Their antimicrobial outcome is owing to obstruction of respiratory enzyme pathways, modification of microbial DNA and the cell wall [5]. Thus, silver and silver ion containing materials are being constantly in use in the field of medical biotechnology [6]. Current studies indicate that silver nanoparticles are very effective as an antimicrobial agent compared to bulk silver or silver ions [7]. The therapeutic effectiveness of silver nanoparticles is numerous folds greater than conventional silver compounds. The nanoparticles employ their antimicrobial property by interacting with the Sulphur comprising proteins existing in the bacterial cell membrane in addition to with the phosphorous containing DNA [7]. Other than that, silver nanoparticles based antimicrobials have added advantages due to their thermal firmness, and Eco friendliness [8]. Hence, the usage of silver based marketable products including topical ointments, bandages, augmentation devices; tissue scaffolds, antimicrobial filters, and gels have amplified for improving public health care [9, 10]. Some commercially existing silver-containing purification arrangements such as Aqua pure, Kinetico and QSI-Nano have shown to eliminate 99.99% pathogens present initially after purification.

Silver nanoparticles (AgNPs) forms composites with polymers such as polyvinyl alcohol, polypyrrole, polyvinylidene fluoride, chitosan, and cellulose as reported in number of research reports [10, 11]. The development of polymer-silver nanocomposites desires that the size of nanoparticles in the polymer matrix be controllable and that their dispersal within the polymer matrix is even [11]. Many prior attempts to form polymer-silver nanocomposites have involved mixing of a nanoparticle solution into the polymerization mixture. These polymer-silver nano composites can have a wide range of application in biomedical fields, such as surgical gloves, antibacterial clothes, sheets and towels, and anti-infectious urinary catheters [12]. They also can be integrated into antiseptic coverings for plastic surgery, shocking wounds, leg sores, skin grafts, cuts, and abrasions. Further, they can be used in numerous household applications such as textiles disinfection in water treatment, food storage containers, home appliances and in medical devices [13].

The idea of the present study was a green synthesis of AgNPs by chemical reduction of silver nitrate using Lupulus amarus extract as a reducing agent and the preparation of AgNPs/Gelatin-PVA nanocomposite film and to study the antimicrobial potential of a silver-Gelatin-PVA nanocomposite system. To the best of our knowledge, this is the first study describing the preparation of silver nanoparticles using Lupulus amarus extract and their composite with Gelatin-PVA along with a cross linker.

2. MATERIALS AND METHODS

2.1 Materials

Silver Nitrate (Sigma), PVA with a molecular weight of 115000 kDa, gelatin (purified) and 25 % Glutaraldehyde Solution was purchased from Merck, Lupulus amarus pallets, Ethanol, D.D Water.

2.2 Green Synthesis of Silver Nanoparticle

2.2.1 Extract Preparation

A 100 mg portion of Lupulus amarus Pallets where crushed into fine powder by a Mortar Pastel and was then added to 100 mL (40% Ethanol) solution and was stirred continuously for 1 h in a Magnetic stirrer and was kept at 4°C for 24 h for maceration. The Mixture is then concentrated using a Rotary Evaporator and finally the concentrated sample was filtered using a Buchner funnel to obtain the final extracted solution (Hops extract).

2.2.2 Synthesis of Silver Nanoparticles

18mL of 0.1mM Solution of silver nitrate was added slowly to 2mL of plant extract with continuous stirring at room temperature for 6 h until the color changes to reddish/brown.

2.3 AgNP –Polymer Film Synthesis

18mL of 0.1mM AgNO3 solution was added to 2mL of Hops Extract and was mixed thoroughly for 6h until silver Nano particles are formed. The formation of silver nanoparticle was confirmed from the change of the color of the solution from pale yellow to dark reddish brown. From the solution 10mL solution is taken and mixed thoroughly with 5mL of 5 % Gelatin solution and 10 mL of 5% PVA solution and the solution was stirred in a Magnetic Stirring Plate (REMI) at a temp of 40-60 °C for 30min. To the homogenized solution, 1mL of 2 % Glutaraldehyde was poured dropwise and the solution was again stirred for another 30mins at a speed of 650 rpm. The final solution was then solvent casted at 40°C for 24 h to obtain the desired membrane. A Gelatin-PVA film was also prepared in similar manner without the nanoparticle.

2.4 Characterization

The Synthesized nanoparticle was characterized using DLS –Zetasizer and using a UV-VIS (carry-60) Spectroscopy to confirm the production of silver nanoparticle. The nanoparticles were scanned from 300 to 800 nm in the UV-VIS spectroscopy.

Morphology of the films and nanoparticles are observed by SEM (Inspect F50 SEM). To image the film samples (surface or cross-sections) were coated with a thin layer of palladium gold alloy after mounting on a double sided carbon tape.

The FTIR spectra of the Nanocomposite films are recorded on Bruker Alpha-Eco ATR. To record the FTIR spectra of films, the samples were completely dried in an oven at 60 °C for 12 h. These samples were read between 500 and 4000 cm^{-1} using ATR mode.

Thermal studies of the films were carried out using Perkin Elmer TGA 4000 System, 100-240V/50-60Hz at a heating rate of 10°C/min and the temperature range was 40-450°C

The wetting property was analyzed by measuring the contact angle of nanocomposite film with respect to Glycerin.

The Shore D hardness of the nanocomposite material was measured using a SHORE-D meter.

2.5 Hemocompatibilty Test

Estimation of Hemocompatibility of the Composites is performed through Hemolysis Studies using fresh human blood, collected in a EDTA tube was diluted with normal saline solution (2 mL blood + 2.5mL normal saline). A standard sample without sharp edges was kept in a centrifuge tube containing 10 mL of normal saline and was kept in an incubator at 37 °C for 30 min. To this was added 0.2 mL of the diluted blood which was then mixed gently and incubated for 60 min. For the positive control, 0.2 mL of diluted blood was taken in 10 mL of 0.1% sodium carbonate solution and for negative control; 0.2 mL of diluted blood was taken in 10 mL of normal saline solution and incubated for 60 min at 37 °C. In a similar way, sample material was incubated for 60 min at 37 °C. After 60 min of incubation, all the test tubes were centrifuged for 5 min at 4000 rpm and the supernatant was carefully removed and transferred to the cuvette for readings at 545 nm wavelength and percentage hemolysis was calculated [14]. Percentage hemolysis is calculated based on average of thee replicates.

$$\%\text{Hemolysis} = \frac{\text{ODtest} - \text{ODnegative}}{\text{ODpositive} - \text{ODnegative}} \times 100$$

1. Highly hemocompatible (<5% hemolysis)

2. Hemocompatible (within 10% hemolysis)

3. Non Hemocompatible (>20% hemolysis)

2.6 Antimicrobial Activity

The antimicrobial activity of the developed nanocomposite films were tested by disc diffusion method using E. coli (gram negative) and S. aureus (gram positive).

For disc diffusion method, the films were cut into a disc shape with 5 mm diameter, sterilized by autoclaving for 30 min at 120 °C, and placed on different cultured agar plates. The plates were incubated overnight at 37°C in an incubator and the inhibition zone was then measured.

2.7 Cytotoxicity experiment

MTT assay was applied to determine the cytotoxicity of prepared sample film on PBMC normal Cell line. The PBMC cells were isolated from blood of a normal human who donated the blood voluntary. Briefly, PBMC cells were dispensed in 24-well culture plates incubated at 37 °C. After 24 h, the nanocomposite film and PBS was introduced to the grown cell. The medium containing only PBS was used as a control. Each group was analyzed in triplicate. After 24 h, 50 µl MTT (5 mg/mL) was added to the medium which was then incubated for another 4 h at 37 °C. The formazan crystals in the cells were solubilized with stock DMSO solution (100 µl/well). The optical density (OD) value was then measured at 590 nm using spectrophotometer. The % viability of is calculated by the following equation [15]:

$$\%\text{Vaible cel} = \frac{\text{OD value of samples}}{\text{OD value of negative control}} \times 100$$

And cytotoxicity was assessed according to cytotoxicity grading criteria as indicated in Table 1.

Table 1: Cytotoxicity grading criteria

Cell Vaibilty (%)	Cytotoxicity grading criteria
100	0 (non-poisonous, qualification)
75–99	1 (light poisonous, qualification)
50–74	2 (moderate poisonous, disqualification)
25–49	3 (severe poisonous, disqualification)
1–24	4 (disqualification)
0	5 (disqualification)

3. RESULTS AND DISCUSSION

At the macro scale, silver always looks like silver. But solutions of silver nanoparticles can have many colors. The main reason behind this color change is the Surface Plasmon resonance [16]. In the silver nanoparticles, electrons oscillate collectively. These oscillations affect how light interacts with the nanoparticles. The specific oscillations depend on the particles' size and shape, so particles of different sizes have different colors. Color change indicates particle size. Solution color gives an approximate idea of the particle size. The color we see is basically an integration of the absorption spectra. Nanoparticle size could be monitored more accurately by taking absorption spectra.

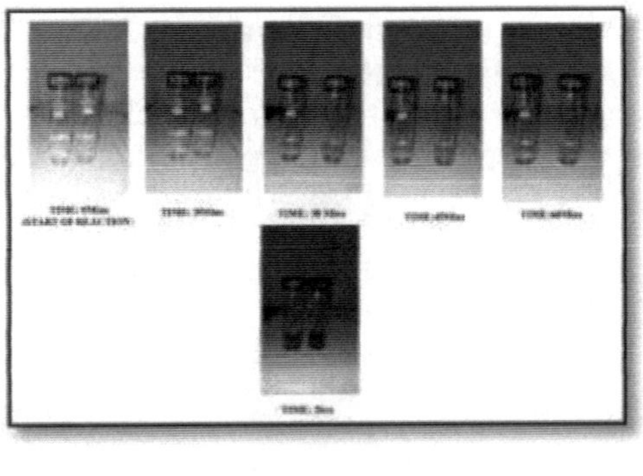

FIGURE 1a : Color change during production of Silver Nanoparticle with time.

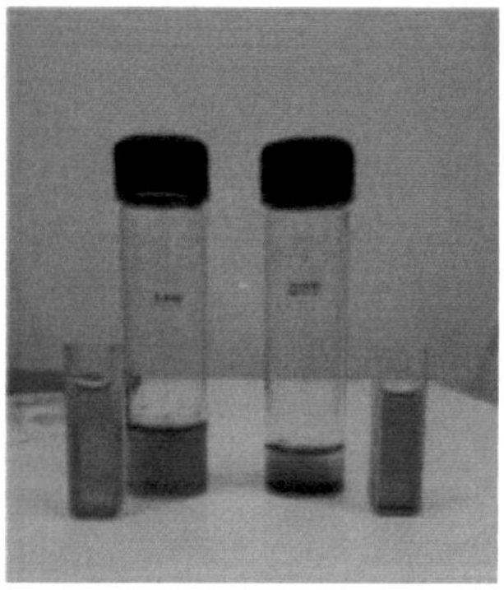

FIGURE 1b : Final color of the solution after 3 h of the reaction with silver nanoparticle precipitation

The absorbance value of reacting mixture measured at the end i.e 3h of reaction in order to verify the concentration of production of silver nanoparticle which is presented in the figure 3. Reduction of silver ions present in the aqueous solution of silver complex during the reaction with the ingredients present in the extract have been evaluated through UV-Vis spectrograph and has been recorded as a function of time with water as reference.

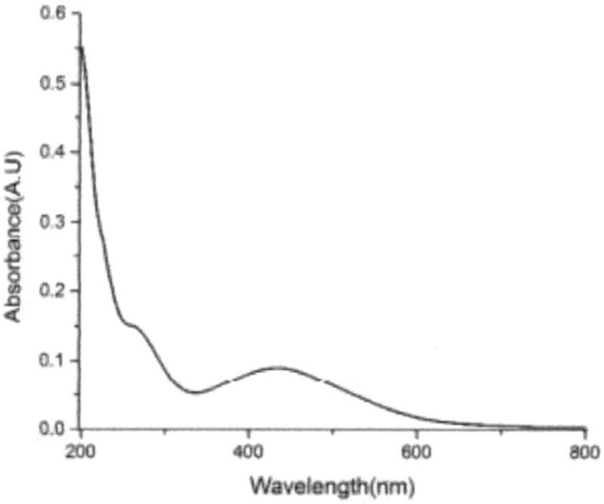

FIGURE 2 : UV Vis spectra of diluted solution of AgNPs

The results of DLS Zetasizer showed very homogenous distribution of AgNPs with average particle size of 86.55 nm as observed from the Figure 4. This denotes to monodisperseity of nanoparticles which provides very high stability of nanoparticles for a long time. In addition, the Poly Dispersity Index (PDI), which is 0.302, specifies high stability and uniformity of the resulting AgNPs. The zeta potential of the synthesized Nanoparticle was also found to be -23.3 mV concluding that the nanoparticle is moderately stable [17].

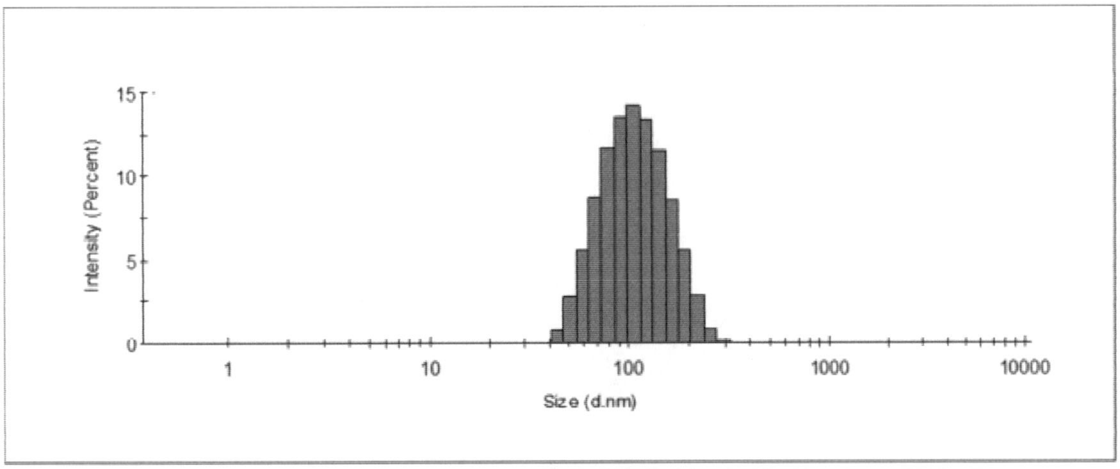

Figure 3: DLS analysis of the nanoparticle

SEM analysis (Figure 4) indicates high-density AgNPs synthesized by the extract. Image indicates moderately spherical and uniform AgNPs were formed with diameter of 15 to 80 nm. The SEM image also illustrated the interactions of hydrogen bond and electrostatic interactions between the bioorganic and phytochemical capping molecules bound to the AgNPs. The nanoparticles were not in direct contact even within the aggregates, demonstrating stabilization of the nanoparticles by the capping agent. The bigger silver particles may be due to the aggregation of the smaller ones.

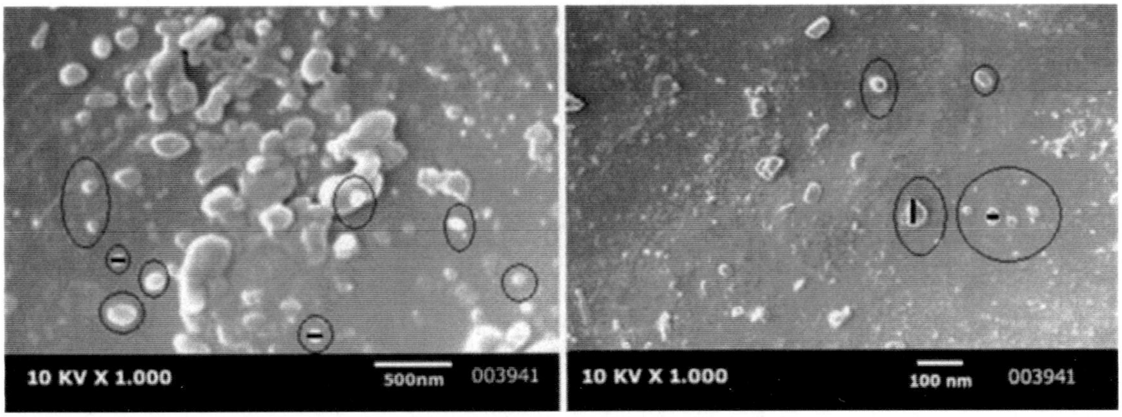

Figure 4: SEM of Nanoparticle

The SEM analysis of the silver nanoparticles loaded Gelatin-PVA films are shown in **Figure 5.** The nanocomposite film has exhibited a dense and uniform plain microstructure. The nanocomposite film also shows the presence of defined nanoparticles in the film. The cross-ectional view of the film indicated a uniform structure with a width of 53.73 μm confirming the formation of thin film.

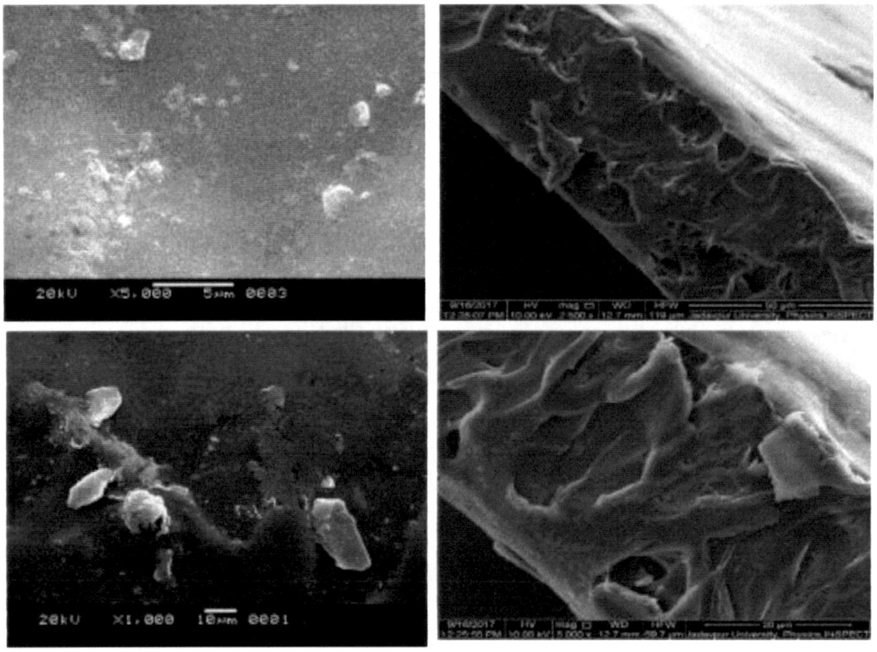

Figure 5: SEM analysis of nanocomposite film; a. surface view, b. cross-sectional view

The interfacial interaction between AgNPS and Gelatin –PVA nanocomposites was confirmed by FT-IR spectra (Figure 6). The infrared spectrum of film features band at 3702 cm^{-1} and 3717 cm^{-1} indicating presence of Amine (N-H stretching) which is for gelatin. Two major peaks at 3397 cm^{-1} and 3311 cm^{-1} indicates the presence of O-H stretching or this is majorly due presence of alcohol which is PVA in this case .Bands at 3084 cm−1 & 2927cm^{-1}due to the stretching of the (C-H) group. The peaks at 1641 cm^{-1}, 1534 cm^{-1}, 1449 cm^{-1}, 1335 cm^{-1}, 1229 cm^{-1}, and 1086 cm^{-1}could be attributed to the presence of aromatic C=C bonds stretching vibrations. In FTIR spectra, the bands at 674 cm^{-1} and 624 cm-1 confirm the presence of a skeletal vibration of C-C groups.

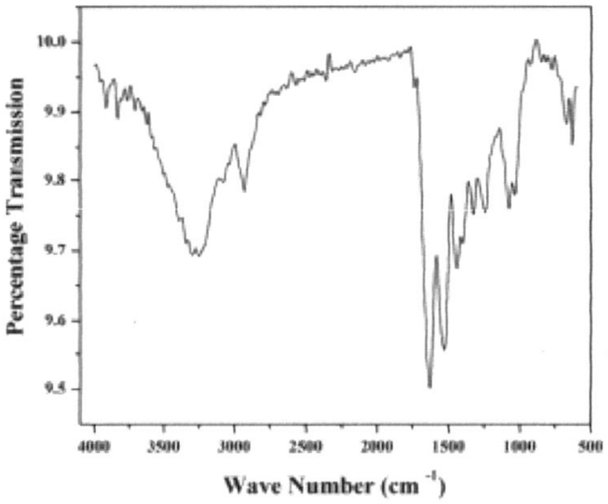

Figure 6: FTIR Spectrum of the Nanocomposite films

at about 160°C, indicating melting of nanocomposite chains and finally the film shows the greatest weight

100°C) observed in the films is due to loss of moisture present in the films. ; The second weight loss starts

IOP Conf. Series: Materials Science and Engineering **410** (2018) 012019 doi:10.1088/1757-899X/410/1/012019

and partial breaking of the molecular structure. Overall the film showed a smooth degradation curve. Thermal degradation starts at around 275 °C. So it will be stable during the sterilization needed before medical application.

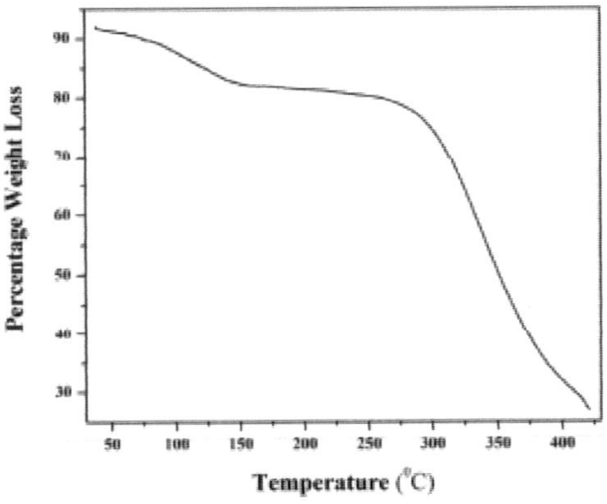

Figure 7: TGA of the Gelatin-PVA Nanocomposite films

The contact angle was measured using glycerin and was found to be 41.4 ° which indicates that the nanocomposite films have a good wetting property or else it can be interpreted that it is highly hydrophilic. As water and Glycerin have near values of surface energy so the contact angle with glycerin and water are likely to same i.e below 90° [18].

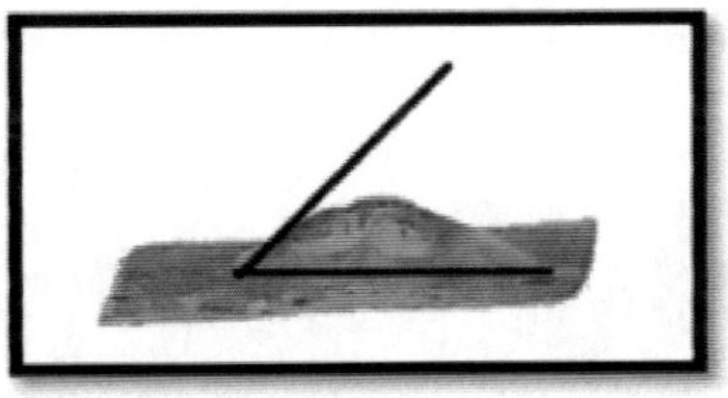

Figure 8: Contact angle with respect to Glycerin

The hardness of the polymer nanocomposite film was tested using Shore D meter and the harness of the film is 6 in the Shore D measurement scale. This indicates that the material is extra soft and can be used for biological purpose [19].

Table 2: Hemocompatibility Test (Human Blood)

Sample	**O.D at 545 nm**	**% Hemolysis**	**Remarks**
+ve Control	0.6629		-
-ve Control	0.0038		-
Gelatin-PVA Silver Nanocomposite	0.0384	5.25%	Hemocompatible

It can be seen from Table 2 that the Nanocomposite films are highly compatible with human blood as the percentage hemolysis id around 5 % which is regarded as hem compatible range. It is reported earlier that Silver nanoparticles alone are highly toxic [20] but here in when it is used as a nanocomposite its toxicity is highly reduced as a total system. But this was only a preliminary test and vigorous biocompatibility tests need to be done with further experiment in animal model.

The antibacterial activity is a demonstration of the release of silver nanoparticles from the polymer network. Silver nanoparticles exhibit relatively large surface area, thus increasing their contact with bacteria. Silver nanoparticles show powerful bactericidal activity by binding with microbial DNA, thereby preventing bacterial replication .The use of Ag-containing Gelatin-PVA films as functional wound dressings is assessed by observing their antimicrobial activity (based on the disc diffusion method) against some common bacteria like *Escherichia coli & Staphylococcus aureus.*

Antimicrobial activity of pure, film and encapsulated silver nanocomposite films, were evaluated from their capacity to inhibit bacterial cultures along with one standard anti-bacterial medicine, Streptomycin (100 μg/mL) available in market was kept as control. The Minimum inhibition concentration of Streptomycin form E. Coli is 32 μg/ml, and that for S. aureus is 42 μg/ml [21]. Figure 9 shows the disc diffusion technique result for different bacterial culture. The experiment contained to sample to confirm the activity.

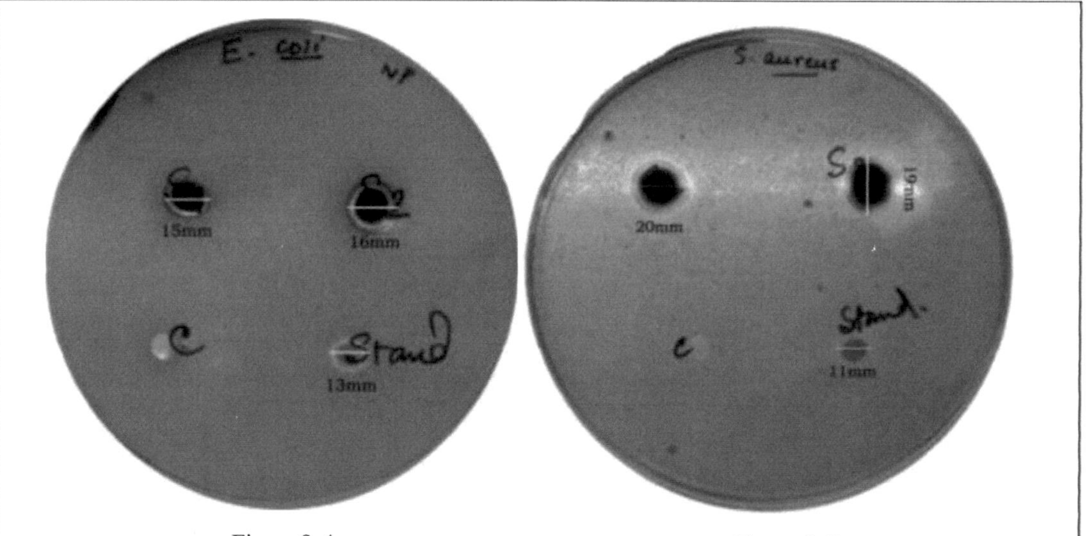

Figure 9-A Figure 9-B

FIGURE 9: Anti-Bacterial Assay using Silver Nanocomposite Gelatine/PVA films

The Ag nanoparticles synthesized and encapsulated in Gelatin /PVA films showed inhibition zone against all test organisms. The Zone of inhibition (ZOI) of AgNP for Escherichia coli (16mm) was greater than that of the standard antibiotic streptomycin (13mm). (Table 3) (Fig 9).

TABLE 3: Zone of Inhibition of AgNPs–Gelatin-PVA films synthesized from Hops extract

Bacterial strains	**inhibition(AgNPsi**	**Zone of inhibition (Standard Drug)**
Staphylococcus aureus	20mm	11mm
Escherichia coli	16mm	13mm

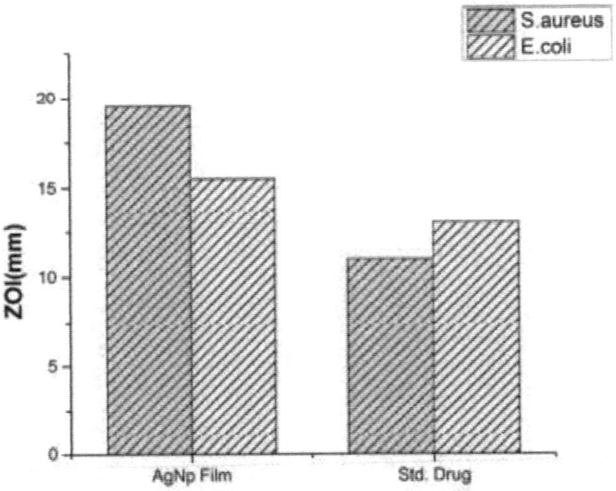

Figure 10: Antibacterial assay Results

The % Cell Viability of PBMC cells on the nanocomposite film was 82.14% (Figure 11) and the cytotoxicity grades of the nanocomposite films was 1 as indicated in Table 1. According to the cytotoxicity grading criteria, the nanocomposite film qualifies as a non-cytotoxic biomaterial.

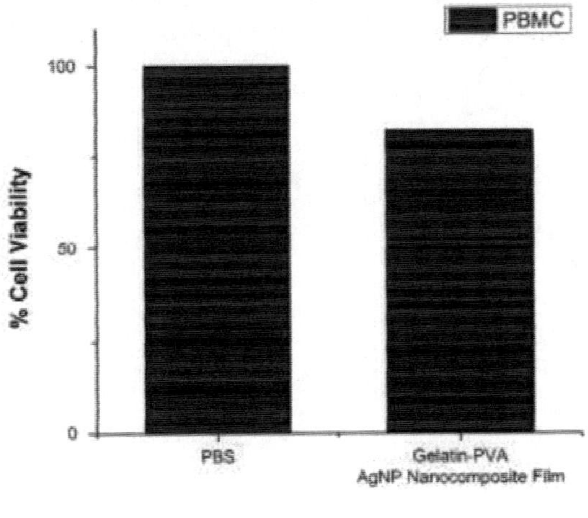

Figure 11: Percentage cell viability assay

4. CONCLUSIONS

The present work demonstrates a simple method for producing novel chitosan-PVA silver nanocomposite films. The developed silver nanocomposite films have exhibited fairly good mechanical strength and superior hemocompatibility properties. Further, the current work demonstrates a promising method to combine silver Nano-composites with a natural compound (Lupulus). The synthesis of Gelatin PVA silver nanocomposite film showed substantial antibacterial activity on both Gram-positive and Gram-negative bacteria. Indicates a potential use of the nanocomposite in the pharmacological, biomedical and industrial fields, such as bandages, wounds dressing, anti-bacterial gloves and dental tools. In addition, the applications include also nutrition food and water storage container as well as wastewater treatment.

5. REFERENCES

[1] E. Amato, Y.A. Diaz-Fernandez, and A. Taglietti, 2011, Langmuir, Vol. 27, pp. 9165.
[2] P. Pallavicini, A. Taglietti, and G. Dacarro, 2010, Journal of Colloid and Interface Science, Vol. 350, pp. 110.
[3] P. Dallas, V.K. Sharma, and R. Zboril, 2011, Advances in Colloid and Interface Science, Vol. 166, pp. 119.
[4] D.G. Yu, M.Y. Teng, W.L. Chou, and M.C. Yang, 2003, Journal of Membrane Science, Vol. 225, pp. 115.
[5] M. Rai, A.Yadav, and A. Gade, 2009, Biotechnology Advances, Vol. 27, pp. 76.
[6] J.H. Crabtree, R.J. Burchette, R.A. Siddiqi, I.T. Huen, L.L. Handott, and A. Fishman, 2003, Peritoneal Dialysis International, Vol. 23, pp.368.
[7] J.S. Kim, E. Kuk, K.N. Yu, J.-H. Kim, S.J. Park, H.J Lee, S.H. Kim, Y.K. Park, Y.H. Park, C.-Y. Hwang, Y.K. Kim, Y.-S. Lee, D.H. Jeong, and M.-H. Cho, 2007, Nanomedicine: Nanotechnology, Biology and Medicine, Vol.3, pp. 95.
[8] R. Kumar and H. Munstedt, 2005, Polymer International, Vol. 54, pp. 1180.
[9] J.J. Castellano, S.M. Shafii, F.Ko, G. Donate, T.E. Wright, R.J. Mannari, W.G. Payne, D.J. Smith, and M.C. Robson, 2007, International Wound Journal, Vol. 4, pp. 114.
[10] M. Mishra, H. Kumar, K. Tripathi, 2008, Digest Journa of Nanomaterials and Biostrctures, Vol. 3, pp. 49.
[11] Y.M. Mohan, T. Premkumar, K. Lee, and K.E. Geckeler, Macromolecular Rapid Communications, Vol. 27, pp. 1346.
[12] T.B. Karchmer, E.T.Giannetta, .A.Muto, B.A. Strain, and B.M. Farr, 2000, Archives of Internal Medicine, Vol. 160, pp. 3294.
[13] C. Marambio-Jones and E. M.V. Hoek, 2010, Journal of Nanoparticle Research, Vol. 12, pp. 1531.
[14] K. Pal and S. Pal, 2006, Material Manufacturing Process, Vol. 21, pp. 325Y.
[15] T. Liu, J. Miao, W. Sheng, 2010, Journal of Zhejiang University Science B, Vol. 11, pp. 10.
[16] A. Vincenzo, O.M. Bakr, and F. Stellacci, 2010, Plasmonics, Vol. 5, pp. 85.
[17] H.K. Ardani, 2017, IOP Conference Series: Materials Science and Engineering, Vol. 188.
[18] Fowkes, M. Frederick M., ed. Contact angle, wettability, and adhesion, American Chemical Society, 1964.
[19] S. Marin, G. Mihail Vlasceanu, R. Elena Tiplea, I. Raluca Bucur, M. Lemnaru, M. Minodora Marin, and A. Mihai Grumezescu, 2015, Current topics in medicinal chemistry, Vol. 15, pp.1596.
[20] R. Hancock, 1960, *Microbiology,* Vol. 23. Pp.179.

IOP Conf. Series: Materials Science and Engineering **410** (2018) 012020 doi:10.1088/1757-899X/410/1/012020

Potential Therapeutic Activity of Bio-Synthesized Silver Nanoparticles as Anticancer and Antimicrobial Agent

Piyali Basak[1], RanabirMajumder[1], Amrita Jasu[1], Samrat Paul[2], Shreya Biswas[2]

School of Bioscience & Engineering, Jadavpur University, Kolkata- 700032, India

Abstract:

Aloevera is one of the oldest medicinal plants and it possesses different types of pharmacological properties including anti-microbial, anti-inflammatory, anti-cancer and antioxidant activity. More over Aloevera contains 75 bioactive compounds such as polyphenol, flavonoids, alkaloids, anthraquinone etc. Such bioactive compounds can be used to reduce silver ions to produce silver nanoparticles. In this study, we are trying to formulate bio-synthesised silver nanoparticle (**b- AgNps**) by green chemistry approach. It is a very simple, efficient and eco-friendly approach for silver nanoparticle synthesis that is formed by reduction of silver nitrate ($AgNO_3$) solution using Aloe vera leaf extract as a reducing agent.

Hydrothermal method was used to prepare b-AgNps using aloe vera leaf extract as both reducing and stabilizing agent. Then b-AgNps were characterized by FESEM, XRD, DLS and UV-VIS spectroscopy. Biological activity of b-AgNps were evaluated by performing cytotoxic test (MTT assay) against breast cancer cell lines and Fibroblast normal cell lines as well as screening antimicrobial activity using agar well diffusion method. The results revealed that there was significant amount of cell growth inhibition against breast cancer cell lines as well as microbes compare to normal healthy fibroblast cell lines.

Keywords: bio-synthesised silver nanoparticle (**b-AgNps**), Alovera, Reducing agent, Breast cancer cell lines.

Corresponding Email. piyalibasak@gmail.com

1. INTRODUCTION

According to the World Health Organization, medicinal plants would be the best source for obtaining a variety of drugs [1]. The gel and dried leaf of Aloe species have been used as medicine since ancient civilizations of the Egyptians, Greeks, Mediterranean people and also for its cosmetic uses [2]. But the attention was drawn to the potential Alo gel after the Second World War when skin burns victims of the nuclear bombs in Japan were successfully treated with aloe gel [3].

The synthesis, delineation of design, and modification of nano-sized materials have found prodigious interest in recent times because of its broad application in medical and biological aspect. Among diverse nanomaterials silver (Ag) was focused with much interest due to its idiosyncratic properties, such as stability, catalytic and antibacterial property. Metallic nanoparticles are produced by various methods, the more common ones being chemical and physical methods. The aforesaid methods produce pure and well-defined nanoparticles, but the chemicals used in the synthesis are toxic, energy consuming, expensive, and unsuitable for various biological applications.
Silver and silver based products like Silver nanoparticles have received attention due to their physical, chemical, and biological properties that attributed to the catalytic activity and

IOP Conf. Series: Materials Science and Engineering **410** (2018) 012020 doi:10.1088/1757-899X/410/1/012020

antimicrobial activity against a wide range of microorganisms like bacteria, fungi, protozoa and recently virus. [4] Beside that, they also found applications in nanobiotechnological research [5,6].

Plant-mediated synthesis of nanoparticles [NPs] is a green chemistry approach that connects nanotechnology with plants. Such plant extracts have reportedly been used in the preparation of AgNPs[7]. Here Aloe extract is used as a reducing agent for the synthesis of biogenic-AgNPs [b-AgNPs]. Water-soluble organic compounds present in the plant materials are mainly responsible for the reduction of silver ions to AgNPs [8]. Besides that aloe vera leaves have been used as medicinal plants as they possess anti-inflammatory activity, anti-arthritic properties, and anti bacterial. They also provide protection against UV ray, promote wound and burn healing, and have antibacterial properties [9,10,11,12]. There are a number of biologically active constituents in aloevera leaves. These include lignin, hemicellulose, pectins which can be used in the reduction of silver ions [13]. As per reports the large enzymes and proteins in aloe vera extract are weakly bound to silver ions and function as a complexing agent. Due to their cost effectivity and environmentally friendly nature coupled with their reducing properties, we selected Aloe vera as the reducing and stabilizing agent to prepare AgNPs and test their anticancer activity.

Biosynthesis of silver nanoparticles (b-AgNPs using *Alo vera* leaf extract, which acts as reducing as well as stabilizing/ capping agent have been shown schematically:

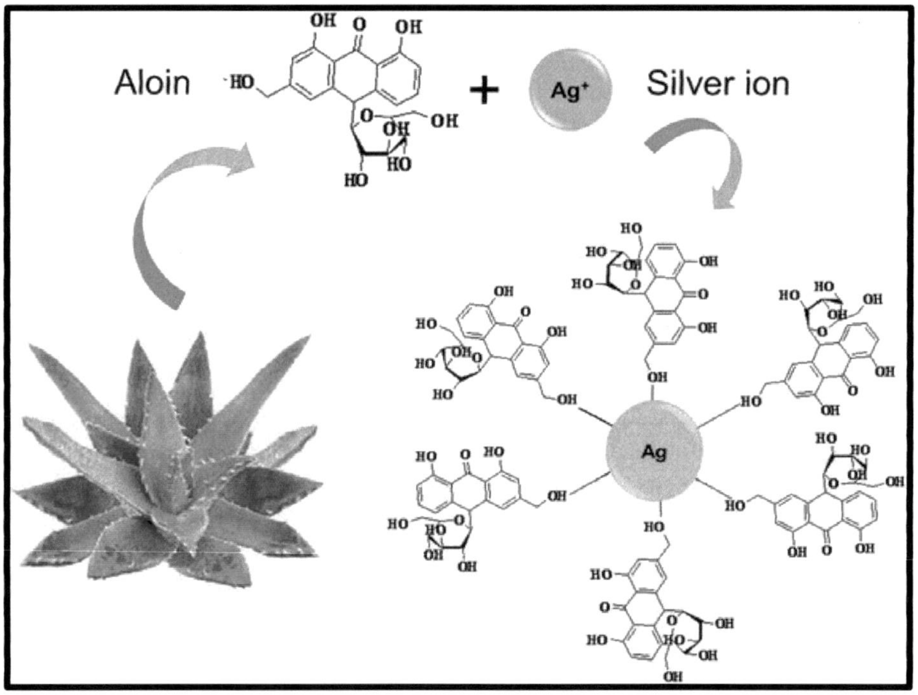

Figure 1 Biosynthesis of silver nanoparticles (b-AgNPs) using *Alo vera* leaf extract as reducing agent

IOP Conf. Series: Materials Science and Engineering **410** (2018) 012020 doi:10.1088/1757-899X/410/1/012020

2. MATERIALS AND METHODS

2.1. Chemicals

Silver nitrate ($AgNO_3$), 3-(4,5-dimethylthiazol-2-yl)-2,5-diphenyltetrazolium bromide (MTT), heat inactivated foetal bovine serum (FBS), minimum essential medium (MEM), glutamine, EDTA and trypsin were purchased from Sigma–Aldrich.

2.2. BioSynthesis of AgNPs

Alovera (AV) whole leaf was obtained from Ramkrishna Mission Ashrama, Narendrapur, Kolkata. Extract (10%) was prepared by mixing 50 g of whole leaf in 500 ml sterile distilled water, kept in boiling water bath for 10 min. The extract was filtered through Whattman filter paper No.1 and stored at 4°C for further studies. Various concentrations of AV-AgNPs were prepared with 1 mM $AgNO_3$ and kept in ambient temperature. Generation of AgNPs was confirmed through periodical measurement using UV–visible spectroscopy in the range between 250-800 nm.

2.3. Characterization of AgNPs

The crystal phase analysis of the AgNPs was conducted using X-ray diffraction (XRD) pattern (Bruker D8 Adance) and was recorded by using Cu-Kα1 radiation (λ of 1.5406Å). The scanning was done in the region of 2θ (from 0° to 80° at 0. 02°/min). The particle size and morphology of the prepared AgNP samples were characterized using field emission electron microscope (FESEM) [Inspect F50, model Q150R with 1400 (VA) power and 50/60 Hz frequency. Size of particle distribution was studies by Zetasizer Nano ZS90. Careful UV–vis spectral analysis was performed by using Varian Cary 50 Bio UV-VIS spectrophotometer to determine the reduction of pure Ag+ ions and thereby the formation and stability of metal nanoparticles in aqueous solution.

2.4. In-vitro anticancer studies of synthesized AgNPs

2.4.1. Cell culture

Breast cancer cell lines (MCF-7) and (T47D) and Fibroblast cells (3T3) were obtained from National Centre for Cell Science (NCCS), Pune, India. The cells were grown as monolayer in MEM, supplemented with 10% FBS, 1% glutamine, and incubated at 37°C in 5% CO_2 atmosphere. Stocks were maintained in 75 cm^2 tissue culture flask.

2.4.2. Cell viability

Cell viability was measured using MTT assay [16]. Cultured MCF-7 cells (1×10^5 cells/mL) were plated on 96 flat-bottom well plates, then cells were exposed to different concentration of AgNPs (1–100 μg/mL) and incubated for 24h at 37°C in 5% CO_2 atmosphere. After incubation, MTT
0.5mg/ml was added to the incubated-cells, then further incubated for another 3 h at 37°C, and 5% CO_2 atmosphere. Thereafter, the formazan crystals were dissolved in 200μl of DMSO and the
absorbance was measured at 570nm in iMark Microplate Absorbance Reader.

2.5 Antibacterial Activity Tests: Well diffusion Method

2.5.1. Bacterial Strains and Culture Media.

The bacterial strains of Gram-negative *Escherichia coli* and Gram-positive *Staphylococcus aureus* were used in this study. Bacterial suspensions were always prepared fresh by growing a single colony overnight at 37°C in a Luria Bertani nutrient broth. The sample turbidity was adjusted to an optical density of 0.1 at 600 (OD 600) before performing the antibacterial experiments. To obtain fresh culture growth, all agar plates were freshly prepared before the antibacterial tests. Cells were added to the luke-warm agar media before plating. These plates were immediately used for the antibacterial activity tests. Cup-Plate method was performed and wells were aseptically dug with a borer.

2.5.2. Determining Zone of Inhibition.
We determined the antibacterial activity of four AV-AgNP solutions (1, 2.5, 5 and 10mM) and each containing the equivalent of 1mM of $AgNO_3$. Simultaneously antibacterial activity of crude Alovera whole leaf extract, dH_2O (kept as control), and standard taken with antibiotic Streptomycin (50µg/ml) were observed. The samples with inoculated agar plates were then incubated for 24h at 37°C. The zone of inhibition (ZOI) was determined as the total diameter (mm) of well with sample diameter plus the halo zone where bacterial growth was inhibited. All measurements were performed in triplicate.

3. RESULTS AND DISCUSSION

3.1 Physico chemical Characterization and stability study of AV-AgNP

In metal nano particles such as in silver, the conduction band and valence band lie in very close proximity where the electrons move freely. These free electrons give rise to a surface plasmon resonance (SPR) absorption band. The absorption peak (SPR) is obtained in the visible range at 410 nm. UV-VIS spectral analysis (Figure 2a) shows the reduction of pure Ag+ ions and thereby the formation and stability of metal nanoparticles in aqueous solution. The crystalline nature of Ag NPs was confirmed by the XRD analysis. The diffraction peaks corresponding to the {111}, {200},
{220} and {311} facets of the fcc crystal structure [14]. The peak corresponding to the {111} plane is more intense than the other planes, suggesting that the {111} plane is the predominant orientation [15]. X-ray diffraction (XRD) pattern (Figure 2b) of AV-AgNPs demonstrates the crystalline nature of silver nanoparticles obtained with 500µL of extract. DLS study (Figure 3) of particle size distribution shows AgNPs with the size range between 85 and 465 nm with more than 75% of AgNPs between 150nm to 300nm and some below 100nm. FESEM micrographs (Figure 4) of different magnifications shows individual and clustered spherical particles as low as 25nm and above. The physicochemical characterization studies carried out using XRD, DLS, SEM, UV- Visible Spectrophotometer, confirmed the presence, distribution and pattern of silver nanoparticles.

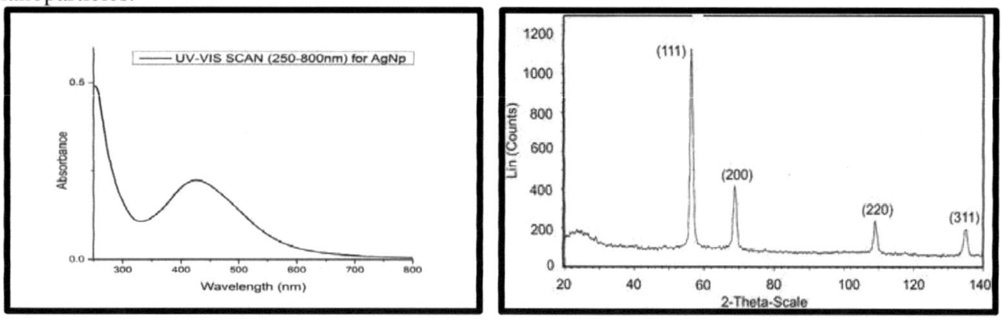

(a) (b)
Figure 2 (a) UV-Vis spectroscopy, (b) XRD pattern of AV-AgNPs hybrids.

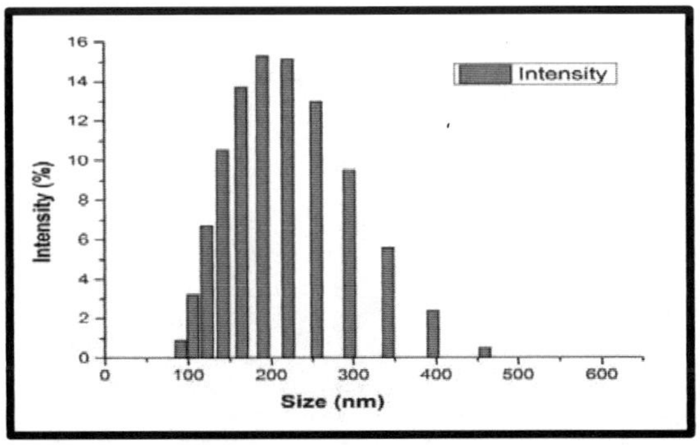

Figure 3 DLS study shows the particle size distribution

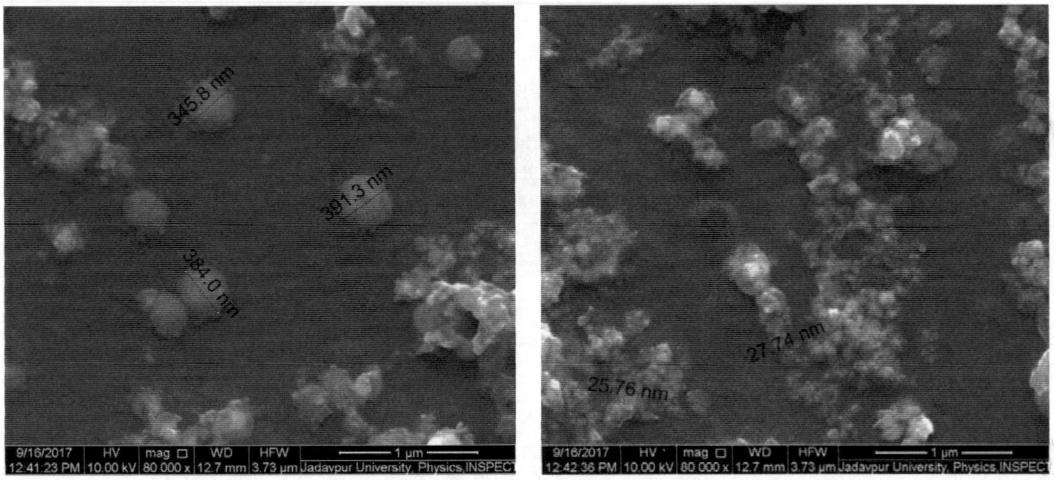

Figure 4 FESEM micrograph of AgNPs at different magnification

3.2. In vitro anticancer studies of synthesized AgNPs

3.2.1. Cell viability

Alo vera crude leaf extract on Fibroblast cell line (3T3) and Breast cancer cell lines MCF-7 and T47D had promoted growth without any cytotxic effect. Whereas Cytotoxicity induced by biogenic AgNPs on MCF-7 and T47D cell lines were found to be higher with increase in concentrations of AgNPs. There was a change in the percentage of cell viability in control, standard with anti-cancer drug Cisplatin(10µg/ml and 20µg/ml) and AgNPs (0, 10, 20, 40, 60, 80 and 100 µg/ml) treated MCF-7 cells. IC_{50} of biogenic AgNPs was found to be 40µg/ml for MCF-7 and 80µg/ml for T47D.

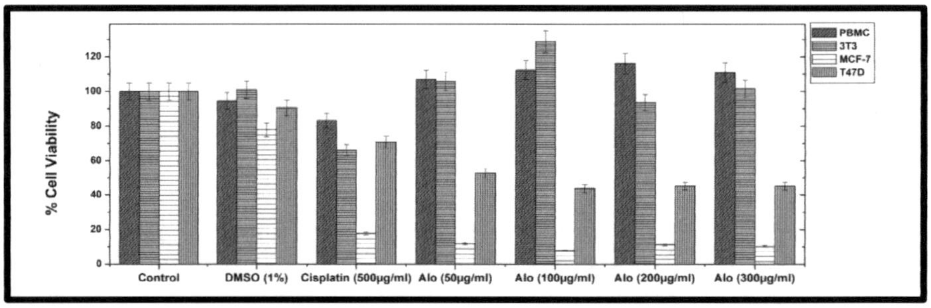

(a)

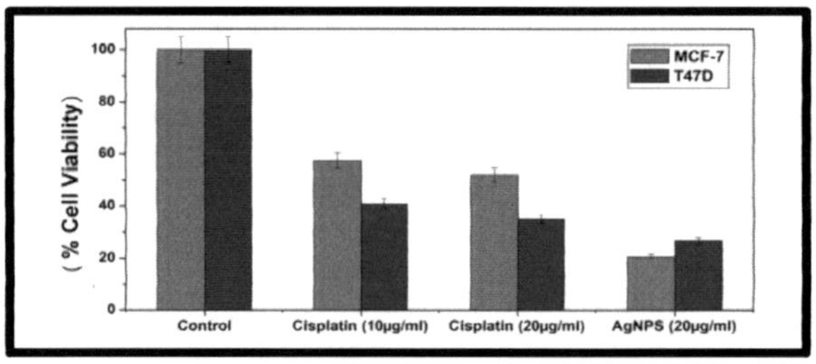

(b)

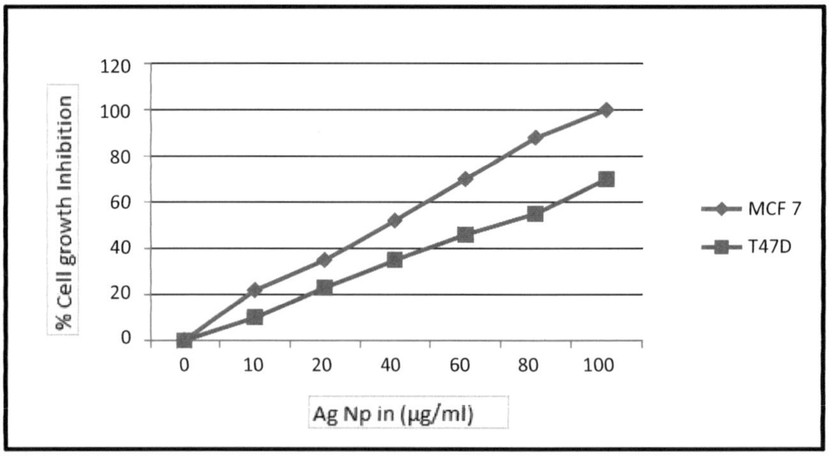

(c)

Figure 5 a. Comparative effect of Aloe vera whole leaf extract on Normal and the Cancer cells, **b.** Cytotoxic effect of biogenic (AV)- AgNPs against MCF-7 & T47D cell line, **c.** Cytotoxicity of AgNPs on MCF-7 and T47D cells: With increased concentration of AgNPs.

3.3 Antimicrobial Study

TheAV- AgNPs synthesized, showed inhibition zone against all test organisms. The ZOI of AgNP for both *Escherichia coli* and *Staphylococcus aureus* were greater than that of the standard antibiotic streptomycin. The ZOI of Gram-positive *S. aureus* increased from 1.3 to 2.1 cm with increasing concentration of AV-AgNPs and Gram Negative *E.coli* showed much prominent effect than *S.aureus* with increasing ZOI from 1.2 to 2.5cm (Table 1).

Table: 1 Comparative study of ZOI of the two strains of Gram-negative *Escherichia coli* and
Gram-positive *Staphylococcus aureus* against the Standard of Streptomycin.

Sample Dose	Zone if Inhibition (cm) of Bacterial strain	
	Escherichia coli	*Staphylococcus aureus*
Standard drug (50µg/ml)	1.1	1.3
1µM	1.2	1.3
2.5 µM	1.5	1.6
5 µM	2.1	1.8
10 µM	2.5	2.1

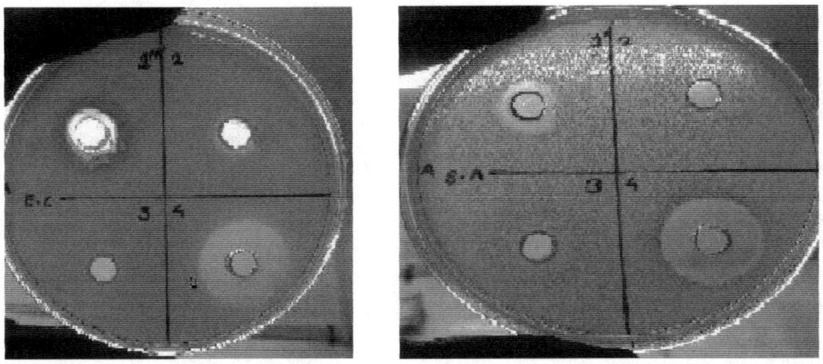

(a) (b)

Figure 6 Plates showing zone of inhibition: (a) *Escherichia coli with 5mM AgNP in 4th quadrant* (b) *Staphylococcus aureus with 5mM AgNPin 4th quadrant.* 1st quadrant with the standard antibiotic drug and 2nd and 3rd quadrant with distilled water and crude Alovera extracts respectively.

4. CONCLUSIONS

Biogenic silver nanoparticles were synthesized using the extract of Alovera as a potent bio reducer. Spherical shaped nanoparticles were observed. Basic experimental evidence shows anticancer property against human breast cancer (MCF-7 and T47D) cell lines. It caused adverse effects of the various cellular components in the cancerous cells. Further in vivo studies and clinical level trails is necessary, to address the formulation of biogenic silver nanoparticles as an eco-friendly and biocompatible alternate to conventional anticancer drugs. Biosynthesized AgNP with Alovera showed clear antibacterial activity toward Gram-negative *E. coli* and Gram-positive *S. aureus*. As all the bacterial species showed similar susceptibility to AV-AgNP, thus can be concluded they show bactericidal action over a wide range of potentially pathogenic bacteria.

IOP Conf. Series: Materials Science and Engineering **410** (2018) 012020 doi:10.1088/1757-899X/410/1/012020

5. ACKNOWLEDGEMENTS

All the authors are thankful to School of Bioscience and Engineering Jadavpur University, Department of Physics, Jadavpur University, Department of Metallurgical and Material Engineering, Jadavpur University.

6. REFERENCES

1. Rates SM. 2001. Plants as source of drugs. Toxicon Vol. 39, pp. 603.
2. Evans WC. 2009. Trease and Evans' Pharmacognosy. Elsevier Health Sciences: London.
3. Reynolds T. 2004. Aloes: The Genus Aloe. CRC press: London.
4. M.I. Sriram, S. Barath, M. Kanth, K. Kalishwaralal, S. Gurunathan, 2010, International Journal Nanomed. Vol. 5, pp. 753.
5. V.K. Sharma, R.A. Yngard, and Y. Lin, 2009, Advances in Colloid and Interface Science, Vol. 145, pp. 83.
6. A.M. Fayaz, K. Balaji, M. Girilal, R. Yadav, P.T. Kalaichelvan, and R. Venketesan, 2010, Nanomedicine: Nanotechnology, Biology, and Medicine, Vol. 6, pp. e103.
7. Sun Q, Cai X, Li J, Zheng M, Chen Z, Yu C-P. 2014. Colloids and Surfaces A: Physicochemical and Engineering Aspects, Vol. 444, pp. 226.
8. Q. Feng, J. Wu, G. Chen, F. Cui, T. Kim, and J. Kim, 2000, Journal of Biomedical Materials Research, Vol. 52, pp. 662.
9. T. Reynolds and A. Dweck, 1999. Journal of Ethnopharmacology, Vol. 68, pp. 3.
10. B. Vazquez, G. Avila, D. Segura, and B. Escalante, 1996, Journal of Ethnopharmacology, Vol. 55, pp. 69.
11. T.H. Emaga, C. Robert, S.N. Ronkart, B. Wathelet and B,M. Paquot, 2008, Bioresource Technology, Vol. 99, pp. 4346.
12. Y. Zhang, X. Cheng, Y. Zhang, X. Xue, and Y. Fu, 2013, Colloids and Surfaces A: Physicochemical and Engineering Aspects, Vol. 423, pp. 63.
13. S.P. Chandran, M. Chaudhary, R. Pasricha, A. Ahmad, M. Sastry, 2006, Biotechnology Progress, Vol. 22, pp. 577.
14. P. Kannan and S. John, 2008, Nanotechnology, Vol. 19, pp. 085602.
15. L. Hart, A.Van der Berg, L. Kuis, H. Van Dijk, R. Labadie, 1989, Planta Medicine, Vol. 55, pp. 509512.

IOP Conf. Series: Materials Science and Engineering **410** (2018) 012021 doi:10.1088/1757-899X/410/1/012021

SYNTHESIS AND CHARACTERISATION OF GELATIN-PVA/HYDROXYAPETITE(HAP) COMPOSITE FOR MEDICAL APPLICATIONS

Piyali Basak[1], *Purbanka Pahari* [1],*Pratik Das*[1],*Nilotpal Das*[1],*Sujan Krishna Samanta*[2] *and Sukumar Roy*[2]

[1] School of Bioscience and Engineering, Jadavpur University, Kolkata - 700032, India

[2] Department of Biomedical Engineering, Netaji Subhash Engineering College, Kolkata -700152, India

Abstract. Biodegradable composite biomaterials play a pivotal role in the healthcare and addressing many challenging issues. Present study reports on the synthesis and characterization of hydroxyapatite (HAP) and gelatine-PVA composite, analyses the incorporation of HAP in gelatine-PVA and investigating their biocompatibility and mechanical properties. The material was prepared in film form by mixing freshly prepared HAP into gelatine-PVA solution and finally drying them to make the composite film. Composite gelatine-PVA were characterized by scanning electron microscopy (SEM), X-ray diffractometry (XRD), Fourier Transform Infrared (FTIR), hydrophilic-hydrophobic nature and evaluated for its strength, hemocompatibility and Cytotoxicity. This composite material as fabricated and characterized, may be used as a target material for bone refurbishment including dental filling patches.

Key words: Composite, Hydroxy apatite, Biocompatibility, Polyvinyl Alcohol
E-mail: purbankapahari36@gmail.com

1. INTRODUCTION

Biomaterials have great importance in the field of Biomedical Engineering. A biomaterial is any material, comprises whole or part of a living structure or biomedical device which performs, augments, or replaces a natural function. The material can be natural or synthetic and includes metals, ceramics and polymers. They mainly are used in the medical field for tissue repair, heart valves etc. In this present experiment, we have studied biomaterials that may be used in bone refurbishment.

Bone is a tissue that has the ability to heal and regenerate itself. Bone refurbishment is a surgical procedure that replaces missing bone in order to repair bone fractures that are extremely complex, pose a significant health risk to the patient, or fail to heal properly. Bone grafting is a surgical procedure that replaces missing bone in order to repair bone fractures that are extremely complex, pose a significant health risk to the patient, or fail to heal properly. Bone grafts may be autologous (bone harvested from the patient's own body), allograft (cadaveric bone usually obtained from a bone bank), or synthetic (often made of hydroxyapatite or other naturally occurring and biocompatible substances) with similar mechanical properties to bone.

For more than 50 years, different materials, such as metals, ceramics, synthetic, and natural polymers have been used for bone repair with different physiological or mechanical characteristics. Manufacturing three dimensional porous scaffolds in recent years increased the hope for the production of scaffolds with closer similarities to bone matrix. Numerous biomaterials have been investigated as scaffolds for bone repair, including natural and synthetic materials. Amongst these, calcium phosphate-based materials, especially bioactive hydroxyapatite (HAP, $Ca_{10}(PO_4)_6(OH)_2$) are widely used because of their chemical and structural similarity to the mineral portion of bone [1, 2]. Natural bone is a three-dimensional composite made up of organic and mineral phases of which collagen is the main organic constituent and HAP is the main mineral one. To synthesize a scaffold that mimics the natural structure of bone, composite systems have been developed, which combine the desired properties of both organic and mineral phases into a single material system. The most widely investigated composites for preparation of tissue engineering scaffolds are polymer/ceramic systems, such as HAP/collagen [3], HAP/chitosan [4], HAP/collagen/poly (lactic acid) [5], HAP/alginate/collagen [6], and HAP/gelatine [7]. Gelatine (GEL), a derivative of collagen, is an attractive component for the purpose of extracellular matrix replacement as it contains several biologic functional groups, which enhance osteoblast adhesion and migration and mineralization.

Polyvinyl alcohol has excellent film forming, emulsifying and adhesive properties. It is also resistant to oil, grease and solvents. It has high tensile strength and flexibility, as well as high oxygen and aroma barrier properties [8].

We have prepared a HAP-Gelatine-PVA film and performed different physical characterisation. We have also evaluated its biocompatible nature to find whether it could be found different medical applications.

2. MATERIALS AND METHODOLOGY

2.1. Chemicals.

HAP[$Ca_{10}(PO_4)_6(OH)_2$], PVA with a molecular weight of 115000 kDa, Gelatin (purified) and 2 % Glutaraldehyde Solution were purchased from Merck.

2.2. Preparation of Hydroxyapatite:

Hydroxyapatite was prepared by reacting calcium hydroxide($Ca(OH)_2$) and phosphoric acid (H_3PO_4). When the reaction was over, the beaker was kept aside for 24 h. It was seen that a white powder like element has settled in the bottom of the beaker and the residual water had come up. The residual water on the top of the beaker was drained out with the help of a sucking tube, and the remaining white coloured product (produced due to the mixture of Calcium Hydroxide and Orthophosphoric acid) was filtered using vacuum pump.

The petridish filled with the end product of the reaction was then dried in an incubator at 70^0c. The dried samples were taken out from the incubator and calcinated in a furnace at 800^0c and heat flow inside the furnace was increased slowly with the help of the knob. It took around 2 h for the temperature of the furnace to reach 800^0c from normal room temperature. The calcination was done for 2 h. The samples were removed from the furnace after 24 h when the temperature slowed down.

IOP Conf. Series: Materials Science and Engineering **410** (2018) 012021 doi:10.1088/1757-899X/410/1/012021

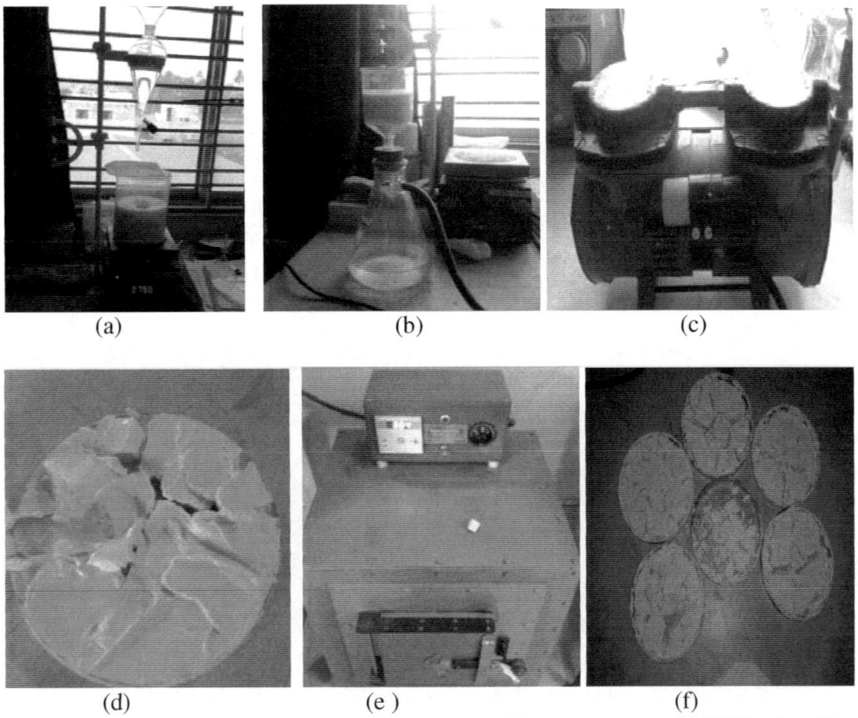

Fig1: (a) HAP preparation; (b) Filtering process; (c) Vacuum pump; (d) HAP after filtering; (e) Furnace machine for drying; (f) HAP after drying.

2.3. *HAP –Polymer Film Synthesis*

200 mg HAP was taken in 10mL acetic acid solution and mixed thoroughly with 5mL of 5 % Gelatin solution and 10 ml of 5% PVA solution and the solution was stirred in a Magnetic Stirring Plate (REMI) at a temp of $40^0 - 60^0 c$ for 30 minute. 1ml of 2 % Glutaraldehyde was poured dropwise into the solution and the solution was again stirred for another 30mins at a speed of 650 rpm. The final solution was then solvent casted at $40^0 c$ for 24 h to obtain the desired membrane. A Gelatin-PVA film was also prepared in similar manner without the hydroxyapatite.

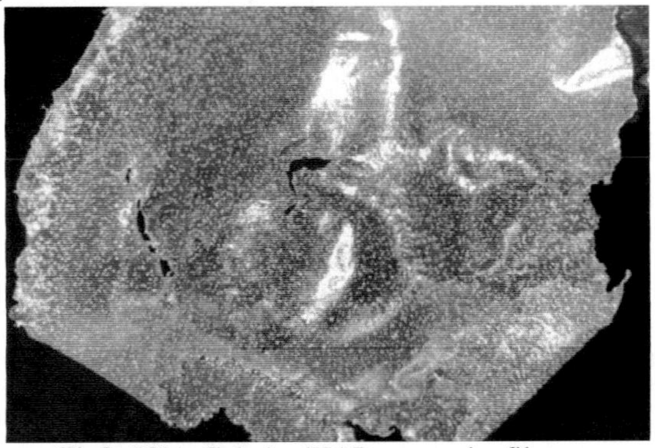

Fig2: PVA-Gelatine/HAP composite film

2.4. Characterization

XRD:
The phase analysis of the HAP powder and composite samples was done by XRD (Model PW 1729, Philips, Holland) using 35 milliamps, and 40 kV current. The mean crystallite size has been calculated using Scherer's equation, that is, $D = 0.9\lambda/\beta \cos \theta$, where D is the average crystallite size in A° , β is the peak broadening of the diffraction line measured at half of its maximum intensity in "radian," λ is the wavelength of X-rays, and θ is the Bragg's diffraction angle[9].

FTIR:
The goal of FTIR is to find different functional group in the present sample [10]. The identification of functional groups in the HAP powder was analyzed by FTIR analysis on IR-Prestige-21 Shinad-ZU at 400-5000 range.

SEM:
Morphology of the films and pure HAP and PVA-Gelatine/HAP composite are observed by Scanning Electron Microscope(SEM) [11] in Inspect F50, model Q150R with 1400 (VA) power and 50/60 Hz frequency. The composite material was coated with gold before study to SEM.

Contact Angle:
Contact angle and biocompatibility is inversely related. When θ=0 then it is highly hydrophilic and biocompatible and when θ>90 it is hydrophobic. The surface property of glycerine and water is quite similar. The wetting property was analysed by measuring the contact angle of HAP-polymer composite film with respect to Glycerine.

.

Mechanical Testing:
Shore durometer is one of several methods used to find the hardness of a material. Higher numbers indicate harder materials; lower numbers indicate softer materials [12]. The Shore D hardness of the composite material was measured using a SHORE-D meter.

2.5. Biocompatibility test

Hemocompatibilty test:

Hemocompatibility of the Composites has been estimated by Hemolysis Studies Fresh human blood, collected in a EDTA tube was diluted with normal saline solution (2 mL blood + 2.5mL normal saline). A standard sample without sharp edges was kept in a centrifuge tube containing 10 mL of normal saline was kept in an incubator at 37^0c for 30 min. To this was added 0.2 mL of the diluted blood which was then mixed gently and incubated for 60 min. For the positive control, 0.2 mL of diluted blood was taken in 10 mL of 0.1% sodium carbonate solution and for negative control, 0.2 mL of diluted blood was taken in 10 mL of normal saline solution and incubated for 60 min at 37^0c. In a similar way, sample material was incubated for 60 min at 37^0c. After 60 min of incubation, all the test tubes were centrifuged for 5 min at 4000 rpm and the supernatant was carefully removed and transferred to the cuvette for readings at 545 nm wavelength and percentageage hemolysis was calculated [14]. %age hemolysis is calculated using percentage equation (1) based on average of three replicates.

IOP Conf. Series: Materials Science and Engineering **410** (2018) 012021 doi:10.1088/1757-899X/410/1/012021

$$\text{Percentage Hemolysis} = [(OD_{test} - OD_{negative}) / (OD_{positive} - OD_{negative})] \times 100 \quad \ldots (1)$$

When the percentage of hemolysis is less than 5% then it is highly hemocompatible, when percentage hemolysis of sample within 10% then the sample is hemocompatible and when greater than 20% then the sample is non hemocompatible.

Cell Viability Study:

The viability test was done by the MTT assay. This colorimetric assay uses reduction of a yellow tetrazolium salt (3-(4,5-dimethylthiazol-2-yl)-2,5-diphenyltetrazolium bromide, or MTT) to measure cellular metabolic activity for cell viability. Viable cells contain NAD (P) H-dependent oxidoreductase enzymes, which reduce the MTT reagent to formazan, an insoluble crystalline product with a deep purple colour. Formazan crystals are then dissolved using a solubilizing solution and absorbance is measured at 595nm. The darker the solution, the greater the number of viable, metabolically active cells.

Absorbance readings from test samples must then be divided by those of the control and multiplied by 100 to give percentage cell viability or proliferation (see formula below). Absorbance values greater than the control indicate cell proliferation, while lower values suggest cell death or inhibition of proliferation.

$$\%\ \text{viable cells} = \frac{(abs_{sample} - abs_{blank})}{(abs_{control} - abs_{blank})} \times 100 \quad \ldots\ldots\ldots\ldots\ldots\ldots (2)$$

3. RESULTS AND DISCUSSION

3.1. Characterisation study

3.1.1. XRD study

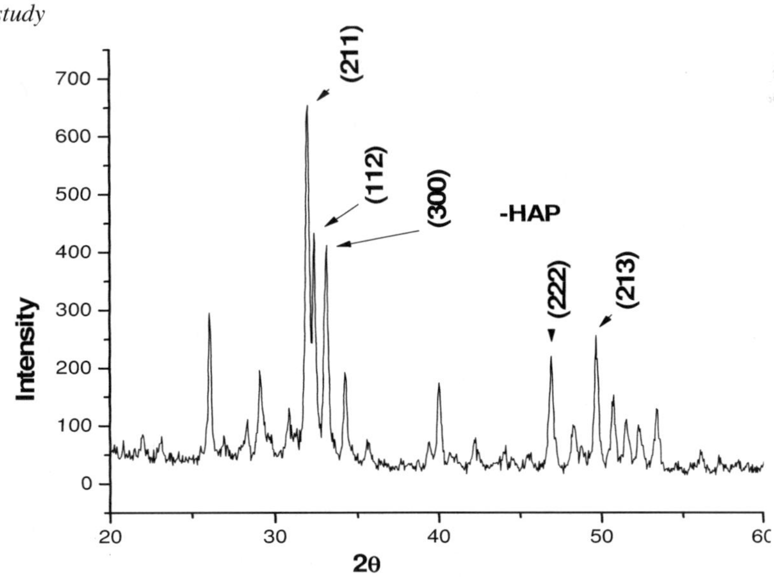

Fig3: XRD curve of Pure HAP

The diffractogram resulting from XRD analysis of the synthesized powder is shown in Fig3. The straight baseline and the sharp peaks crystal of this diffractogram in 211, 112, 300, 222

IOP Conf. Series: Materials Science and Engineering **410** (2018) 012021 doi:10.1088/1757-899X/410/1/012021

and 213 was measured in $30^0 - 60^0$ range which confirms that the product is well crystallized.

3.1.2. FTIR study

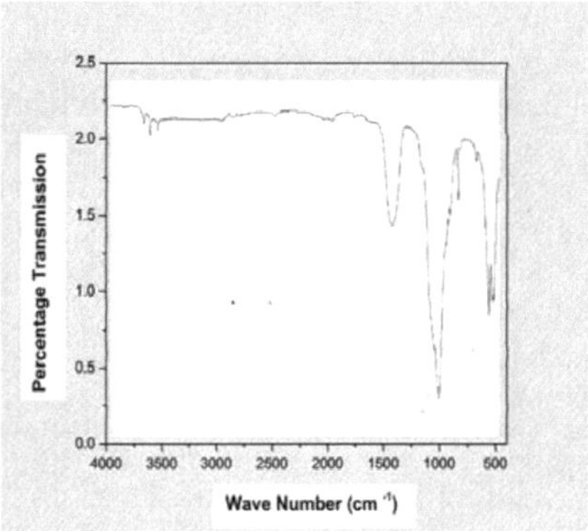

(a)

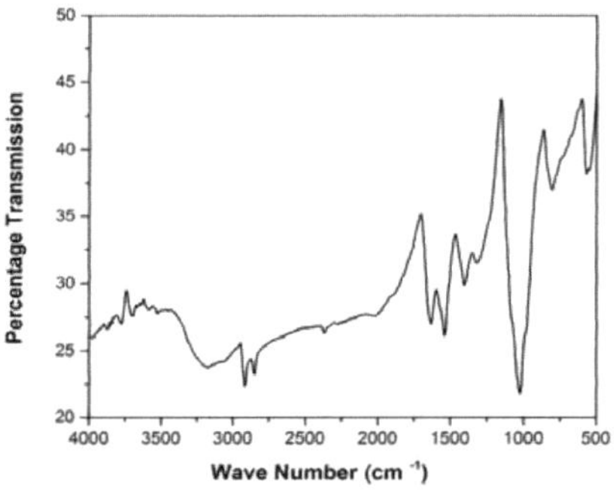

(b)

Fig4: (a) FTIR study of pure HAP (b) FTIR study of PVA-Gelatine/HAP composite

In Fig4 (a) FTIR Spectra clearly confirms the presence of PO_4^{3-}, OH^- and CO^{3-} group in the developed products. The peak observed around 3630 cm^{-1} is due to the presence of –OH bond. This peak is mainly due to O-H stretching vibration in HAP. The peak at 1550 cm^{-1} for $CO3^-$ and the peak at 1100 cm^{-1} is associated with the stretching modes of the P-O bonds of HAP. The double peak at around 600 cm^{-1} is due to the bending modes of P-O bond in phosphate.

The interfacial interaction between PVA-Gelatine/HAP composites was confirmed by FT-IR spectra in Fig4 (b). The peak at 3650 indicates the presence of O-H stretching or this is majorly due presence of alcohol which is PVA in this case. Instead sharp a broad peak also indicative of H-bonded OH of PVA. Band at 1100 cm−1 due to the stretching of the P-O bonds. The peaks at 1500 and 1600 could be attributed to the presence of aromatic C=C bonds stretching vibrations.

.

3.1.3.SEM:

Fig5: SEM study of pure HAP in 5000 magnification

IOP Conf. Series: Materials Science and Engineering **410** (2018) 012021 doi:10.1088/1757-899X/410/1/012021

The analysis of SEM micrograph for HAP above shows in Fig5 fine agglomerates that are of irregular shape, oval and spherical-pellet-like shapes. Particle are formed from 0.4 μm to 1.6 μm in size.

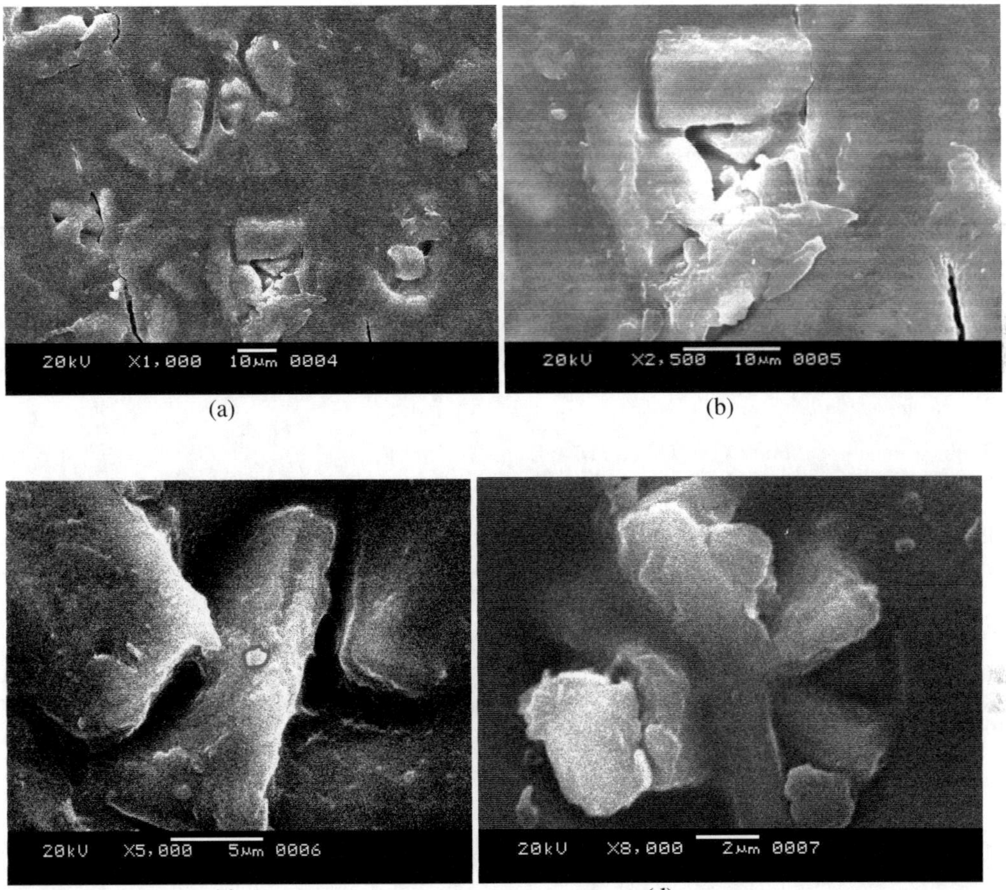

(c) (d)

Fig6: SEM study of PVA-Gelatin/ HAP composite in (a)1000 magnification; (b) 2500 magnification; (c) 5000 magnification and (d) 8000 magnification.

Fig6 shows the SEM micrographs of the composite sample in different magnification. Presence of HAP within the polymer matrix is clearly indicated through the SEM. As there is two type of polymer. Formation of different layer of polymer is also reflected in the SEM. Beside that presence spherical HAP particle size is around 0.9 μm to 0.14 μm which is nearly similar to that of only HAP as indicated in Fig5.

3.1.4. Contact Angle:

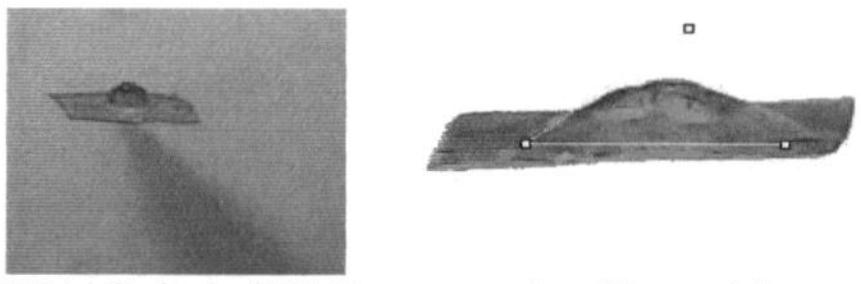

The contact angle of PVA-Gelatin/HAP composite film with respect to Glycerine measured 44^0 which is indicating the biocompatible property of the composite material.

3.1.5. Mechanical Testing

The hardness of the HAP-polymer composite film was tested using Shore D meter and the hardness of the film is 14 in the Shore D measurement scale which implies the composite material is very soft in nature.

3.2. Biocompatibility test

3.2.1. Hemocompatibilty test

Table 1: Hemocompatibility Test (Human Blood)

Sample	O.D at 545 nm	% Hemolysis	Remarks
+ve Control	0.6629		-
-ve Control	0.0038		-
Gelatin-PVA /HAP composite	0.0088	0.759%	Hemocompatible

Table 1 shows that the HAP-Polymer composite films are highly compatible with human blood.

3.2.2. *Cell Viability Study*

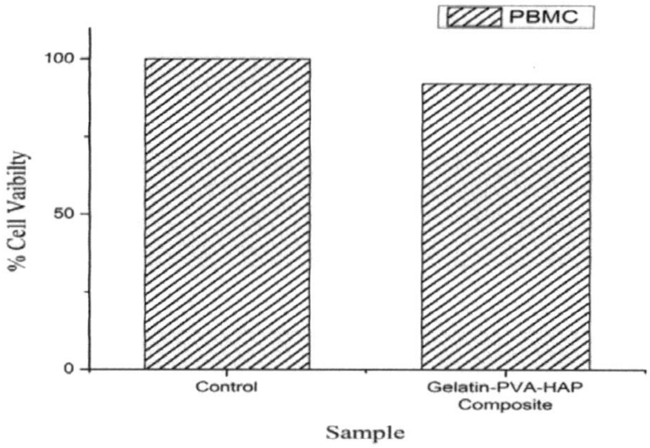

Fig7: Cell viability test on HAP-Polymer composite film, Pbs was taken as control

It can be seen from Fig7 that the HAP-Polymer composite films are highly cell viable (above 90.14%) so these composite materials are highly biocompatible and has very less toxic effect.

4. CONCLUSIONS

In our work, we have seen XRD analyses clearly confirm the synthesized powder was nearly a pure HAP. All the typical absorption characteristics of HAP were also observed in FTIR analysis of the synthesized powder and the resulting scaffold. Hemolysis test and viability confirms that the developed composite is highly hemocompatible whereas contact angle property confirm its hydrophilic nature. From the hardness testing we knew this composite is very soft material and mtt assay confirms its biocompatible nature. So we can conclude it may be has application in vivo. Therefore, the developed Gelatine-PVA /HAP composite may have applications further evaluated in bone tissue engineering applications and in bone refurbishment.

5. REFERENCES

[1] M. Kikuchi, S. Itoh, S. Ichinose, K. Shinomiya, and J. Tanaka, 2001, Biomaterials, Vol. 22 pp. 1705.
[2] S.-C. Liou, S.-Y. Chen, and D.-M. Liu, 2003, Biomaterials, Vol. 24, pp. 3981.
[3] S. Itoh, M. Kikuchi, Y. Koyama, H.N. Matumoto, K. Takakuda, K. Shinomiya, and J. Tanaka, 2005, Materials Engineering, Vol. 15, pp. 29.
[4] I. Yamaguchi, K. Tokuchi, H. Fukuzaki, Y. Koyama, K. Takakuda, H. Monma, and J. Tanaka, 2001, Journal Biomedical Materials Research Part A, Vol. 55, pp. 20.
[5] S.S. Liao, F.Z. Cui, and Y. Zhu, 2004, Journal Bioactive and Compatible Polymer, Vol. 19, pp. 117.
[6] S.M. Zhang, F.Z. Cui, S.S. Liao,Y. Zhu, and L. Han, 2003, Journal of Materials Science and Materials Medicine, Vol. 14, pp. 641.
[7] H.-W. Kim, J.C. Knowles, and H.-E. Kim, 2005, Journal of Biomedicine Materials Ressearch part A, Vol. 72, pp. 136.`
[8] J. Fromageau, E. Brusseau, D. Vray, G. Gimenez, and P. Delachartre, 2003, IEEE Transection Ultrason. Ferroelectr. Frequecy Control, Vol. 50, pp. 1318.
[9] P.P. Ewald, 1962, *50 Years of X-ray Diffraction* (Utrecht)
[10] R. Salzer, P.R. Griffiths, and J.A. de, 2008, *Analitical and Bioanalitical Chemistry,* Vol. 391, pp. 2379.
[11] K.C.A. Smith and C.W. Oatley, 1955, British Journal of Applied Physics, Vol. 6, pp. 391.
[12] A.W. Mix and A.J. Giacomin, 2011, *Journal of Testing and Evalution,* Vol. 39, pp. 696.

Predictive modelling and analysis of surface roughness in CNC milling of green alumina using response surface method and genetic algorithm

Raja Das [1], Sushree Swati Mohanty [2], Muktikanta Panigrahi [3], Saralasrita Mohanty [4,5]

[1] School of Advanced Sciences, VIT University, Vellore, TN-632014, India
[2] Schools of Life Science, Ravenshaw University, Cuttack-753003
[3] Metallurgical Engineering Department, Gandhi Institute of Engineering and Technology (GIET), Gunupur, Rayagada, Odisha, PIN-765022, India
[4] School of Physical Sciences, National Institute of Science Education and Research, HBNI, Jatni - 752050, India

ABSTRACT

Modelling and optimization of machining parameters are essential in Computer Numerical Control (CNC) milling process. The objective of current study is to develop a functional relationship between various factors and responses of CNC machined alumina green ceramic compact. As, ceramic material is notch sensitive in nature, the measurement of average surface roughness (Ra) is vital as it influences the quality and performance of the finished product. In this context, optimization of surface roughness is of maximum importance in manufacturing sectors. To accomplish the required optimal levels of surface quality, the proper selection of machining parameters in CNC milling is highly needed. In this study, four significant machining parameters including spindle speed, XY speed, Z speed and depth of cut in CNC milling process have been selected and along with various combination experiments were conducted. A mathematical regression model was developed to predict the average surface roughness in CNC milling machined surface of alumina based green ceramic compact. The developed model was validated with the new experimental data. Further, the model was coupled with Genetic Algorithm (GA) technique, to predict the optimum possible surface roughness. The results demonstrate the potential to improve the efficacy of production and quality of the finished product as well.

Keywords: CNC Milling, Surface Roughness, Response Surface Method, Genetic Algorithm

Corresponding author -e-mail: saralasrita@niser.ac.in

1. INTRODUCTION

Advanced ceramics is increasingly popular in several engineering applications ranging from automotive and electronics to aerospace fields due to their superior properties including high corrosion resistance, wear resistance, chemical stability, compressive strength and stiffness at high-temperature [1, 2, 3]. Different new technologies such as Computer Numerical Control (CNC) machine tool, lithography, ultrasonic, beam etching and laser machining are adapted for mould-free fabrication of advanced ceramics for various applications [4, 5, 6, 7]. However, sintered state ceramic machining create barrier for economically and efficiently productions due to its brittle and high hardness value and thus imposing high machining cost and hindering further applications [8]. In this context, green state (unsintered) machining of ceramics is considered as a potential alternative for manufacturing of high quality products due to high material removal rates, consumption of lower energy and lesser tool wear [9, 10, 11]. CNC machining of green ceramics is one of the machining methods for fabrication of miniaturized components using diamond embedded pointed tool [12, 13].

Apart from material speciation and dimensions tolerance, the quality of the machined ceramic surface is greatly influences the performance, longevity, reliability, surface texture and manufacturing cost of the finished ceramic part either directly or indirectly. The finer irregularities of the machined surface texture leads to affect other functional features of the final product such as work piece rejection, friction, fracture etc. In CNC machining of green ceramics operations, several factors such as uncontrollable factors (cutting tools material properties and geometry) and controllable factors (cutting speed, depth of cut, spindle speed and feed rate) significantly influence the final surface finish of the ceramic part [14]. Generally, different machining parameters are chosen based upon the prior experience of the person who performed the cutting operation followed by optimization of the process parameters for obtaining better surface finish.

However, this conventional trial and error method is not only a tedious task but also time consuming and ultimately provide high cost. To avoid time and money consuming experiments, different statistical techniques are employed to evaluate the surface finish parameters and optimizing the controllable factors in order to obtain desired level of surface roughness of the final ceramic components. Among different surface finish parameters, average surface roughness (Ra) is widely used as surface parameters in industry. Several studies have been performed in the past few years in order to predict the surface roughness value using statistical tool [15, 16, 17]. Previously, Suresh et al. developed a mathematical model for surface roughness (Ra) prediction through response surface method (RSM) while machining mild steel using carbide-coated tool [18]. In this study, a genetic algorithm (GA) was used and results were compared with RSM to validate the objective function. Further, in order to obtain maximum material removal rate at minimum surface roughness and cutting forces, Prajina implemented RSM in CNC end milling operation [19].

IOP Conf. Series: Materials Science and Engineering **410** (2018) 012022 doi:10.1088/1757-899X/410/1/012022

Routara et al. predicted surface quality through optimizing the different machining parameters including depth of cut, spindle speed and cutting speed of CNC end milling by using Taguchi method [20]. Oktem et al. combined RSM along with GA and neural network to analyze the optimum cutting conditions for obtaining minimum average roughness in aluminium and plastic mold parts using CNC milling machine [21].

However, from the literature survey it can be realized that no one reported the result for prediction the surface roughness of machinable green alumina through CNC milling using diamond embedded tool [12]. The main objective in this work is to investigate the best combination of machining parameters of CNC milling of green alumina compacts using diamond-embedded tool to achieve desired level of surface roughness. In this context, RSM has been employed as a mathematical model using different combinations of cutting parameters from the experimental data (Table 1) and developed model is tested with the experimental data for reliable predictions of surface roughness and optimize machining performances.

Table 1. Experimental Values

	Levels			
Variables	1	2	3	4
XY speed(mm/s)	4	6	8	10
Z speed(mm/s)	1	3	5	7
Spindle speed(rpm)	8000	10000	11000	12000
Depth of cut(mm)	0.1	0.3	0.5	0.7

2. METHODOLOGY

2.1. FABRICATION OF GREEN ALUMINA COMPACTS

The process for fabrication of green alumina compacts with reasonable strength has been reported by Mohanty et al [22]. In brief, the aqueous slurry consists of highly loaded alumina powder (RG 4000, Almatis (Germany), alongwith polymaleic acid (PMA, Aquapharm, India) used as dispersant. The volumetric % of aqueous alumina slurry was 55% in which PMA of 3.8 mg per gram of alumina powder (optimized) was used. Further, in the mixture, sucrose and ovalbumin were used as binders and the binder amount was optimized by rheological studies. Finally, according to the optimized data, 10 volume % ovalbumin, 3 wt% sucrose were added to the aqueous solution. Zirconia balls (3mm diameter, Jyoti Ceramics, India) which functions as milling media were added to prepare the constituents of 55% volume alumina loaded slurry. The homogenous slurry with good owability was produced after 24hrs of ball milling. And through sieving, the milling media has been removed from slurry. After this process, an anti-foaming agent 1-octanol (1ml/100ml of slurry, Merck, India) was added to the slurry. The function of anti-foaming agent is to remove the air bubbles. Distinctive rectangular silicon elastic molds were utilized for slurry casting. The slurry dried at 40C to avoid warpage connected with drying under controlled humid conditions. From the mold, the dried alumina were removed and the rectangular green alumina compacts was ready for CNC machining.

2.2. MACHINING OPERATION

For CNC machining of green alumina rectangular compacts, a bench-top CNC milling machine (MDX 540, Roland DG Ltd., Japan) was used. Further, a conically pointed end diamond embedded tool with mesh size of 625 was used to assess surface roughness of the green alumina compacts. Before machining operation, the CNC machine table is mounted with vacuum dried green alumina. The green alumina samples were machined by CNC milling machine using diamond embedded tools. Preceding machining "stl" configuration of 3D pictures of rectangular block was prepared using solidworks to machine alumina compacts through CNC machining. The conically pointed end diamond embedded tool was mounted at the tool post of the CNC machine. The green alumina compact was machined by raster pattern on the top surface of the green alumina compacts using different machining parameters [12, 23].

2.3. MEASUREMENT OF SURFACE ROUGHNESS

The surface roughness of the machined green alumina compacts was measured using surface profilometry (Talysurf i60/i120/i200-Inductive Systems, Taylor Hobson Limited, Leicester, England). The mean Arithmetic roughness (Ra) was measured by surface profilometer at 1 mm interval of each sample along and perpendicular direction of machining using contact type stylus profilometer with a 1mm tip radius.

IOP Conf. Series: Materials Science and Engineering **410** (2018) 012022 doi:10.1088/1757-899X/410/1/012022

3. REGRESSION MODEL DEVELOPMENT USING RESPONSE SURFACE METHOD (RSM)

In the recent paradigm, RSM method is very often employed to establish the relationship between different input parameters with output one. The rationale of the present work is to predict the average surface roughness (Ra) of CNC milled green alumina compacts by making a relationship between the different input process parameters of CNC milling with the output surface finish through RSM. In this aspect, a second order polynomial response surface mathematical model was undertaken as shown in the Equation (1) to study the influence of input machining parameters of CNC milling of green alumina compact on its surface finish.

$$Ra = a_0 + a_1x_1 + a_2x_2 + a_3x_3 + a_4x_4 + a_5x^2_1 + a_6x^2_2 + a_7x^2_3 + a_8x^2_4 + a_9x_1x_2 + a_{10}x_1x_3 + a_{11}x_1x_4 + a_{12}x_2x_3 + a_{13}x_2x_4 + a_{14}x_3x_4 \quad (1)$$

where,

x_1 = XY speed(mm/s), x_2 = Z speed(mm/s), x_3 = Spindle speed(rpm) and x_4 = Depth of cut(mm).

The coefficient a_0 is the constant term; the coefficients a_1,a_2, a_3 and a_4 are the linear terms; the coefficients a_5,a_6, a_7 and a_8 are the quadratic terms; the coefficients a_9,a_{10},a_{11},a_{12},a_{13} and a_{14} are the interaction terms. The coefficients of regression model were estimated from the experimental results and the model is developed.

4. GENETIC ALGORITHM (GA)

Genetic algorithm was employed to acquire the optimum machining parameters for output surface roughness by using the numerous combinations of the input parameters. In the present study MATLAB code was used to perform the computational algorithm. The simplest Genetic algorithms (GAs) consist of different operators including reproduction, crossover, and mutation. The reproduction is characterized by copying the best individuals from one generation to the next. In this context, the best solution is monotonically improving from one generation to the next. For the crossover operator, the chosen parents are submitted for generating one or two children. The crossover is performed with an allotted probability, which is usually rather high. It is noteworthy that the crossover is carried out when a number, sampled at random, is inferior to the probability. The genetic mutation establishes diversity in the population through an occasional random replacement of the individuals. Using an assigned probability, the mutation is carried out. A random number is employed to determine if a new individual will be produced to substitute the one generated by crossover. The mutation procedure involves substituting one of the decision variable values of an individual, without affecting the remaining variables. The changed variable is randomly chosen, and its new value is taken by randomly sampling within its specific range. The regular genetic algorithm is stated as given below.

BEGIN
INITIALIZE population with a random candidate solution EVALUATE each candidate;
REPEAT UNTIL (terminate conditions) is satisfied DO
1. **SELECT** parents;
2. **RECOMBINE** pairs of parents;
3. **MUTATE** the resulting o spring;
4. **SELECT** individuals or the next generation;
END

5. OPTIMIZATION OF CNC MILLING PARAMETERS

In view of the above, GA was employed as an optimization technique to address a bound-constrained optimization problem. The regression model developed by response surface methodology was employed as an objective function and the upper and lower bound parameters are identified by conducting experiments. The problem can be formulated as given below.

Minimize

$$Ra = f(x_1, x_2, x_3, x_4)$$

Where,

$$1 \le x_1 \le 4$$

$$4 \le x_2 \le 10$$

$$8000 \le x_3 \le 1200$$

$$0.1 \le x_4 \le 0.7$$

6. RESULTS AND DISCUSSION

RSM was developed by using the average surface roughness (Ra) of green alumina compacts obtained from the experimental results. For Ra in perpendicular and parallel tool movement path directions, ANOVA was used to confirm the value of the regression model and model coefficients. Moreover, it was employed for analyzing the null hypothesis of the experimental data with a confidence level of 95%. It is to be noted that if the p-value 0.05 for the F -statistic, H_0 is true and treatments have statistically no effect. However, If the p-value 0.05, it is to be concluded that H is true and the treatments have a statistically significant effect. The average surface roughness value (Ra) that acquired from the experimental data was compared with the predicted value obtained from the model. Table 2 and Table 4 are ANOVA outline tables of the terms in the model, to decide whether to eliminate or fail null hypothesis for average surface roughness in the perpendicular and parallel tool motion directions, respectively. Further, to design the regression model, the

Minitab software was employed.

It was observed that p-value of six terms are less than 0.05 for both the directions, hence they are significant in the regression models. Table 3 and Table 5 demonstrates the ANOVA analysis of the models, and the column illustrating the degrees of freedom (DF), the sequential sums of squares (Seq:SS), the adjusted sums of squares (AdjSS), the adjusted mean squares (AdjMS), the F -statistics from the adjusted mean squares, and its p-value. The sequential sum of squares is the added sums of squares given that prior terms are in the model, which depends on the model order.

Table 2. ANOVA summary of surface roughness in perpendicular direction

Term	Coeff .	SE Coeff .	T	P
constant	0.544944	0.02675	20.375	0
x_1	0.063875	0.02057	3.105	0.005
x_2	0.016525	0.02057	0.803	0.43
x_3	0.199292	0.01534	12.996	0
x_4	-0.01275	0.02057	-0.62	0.541
x_1^2	0.072131	0.0308	2.342	0.028
x_3^2	-0.015208	0.02656	-0.573	0.572
x_1x_2	-0.049802	0.02155	-2.31	0.03
x_1x_3	0.025838	0.0252	1.025	0.315
x_1x_4	-0.080341	0.02155	-3.727	0.001
x_2x_3	0.00465	0.0252	0.185	0.855
x_3x_4	-0.030488	0.0252	-1.21	0.238

The adjusted sums of squares are the sums of squares given that all other terms are in the model that does not depend upon the model order. It can be observed in this table, the p-value of regression model is less than 0.05, hence, the Ra fitting the regression model with the linear, and square terms are significant at the level of 95%. The terms x^2_2, x^2_4, and x_2x_4 are not included in the 0.05 for both the perpendicular and parallel directions. This way the simplified truncated models is shown in the following equations (2) and (3).

Table 3. Variance analysis of surface roughness in perpendicular direction

Source	DF	Seq. SS	Adj. SS	Adj. MS	F	P
Regression	11	1.1541	1.1541	0.104918	18.59	0
Linear	4	1.04955	1.04955	0.262387	46.49	0
Square	2	0.00773	0.03281	0.016404	2.91	0.074
Interaction	5	0.09682	0.09682	0.019365	3.43	0.018
Residual error	24	0.13546	0.13546	0.005644		
Total	35	1.28956				

IOP Conf. Series: Materials Science and Engineering **410** (2018) 012022 doi:10.1088/1757-899X/410/1/012022

Table 4. ANOVA summary of surface roughness in parallel direction

Term	Coeff .	SE Coeff .	T	P
constant	0.210528	0.014601	14.418	0
x_1	0.01715	0.011233	1.527	0.14
x_2	0.03695	0.011233	3.29	0.003
x_3	0.082792	0.008372	9.889	0
x_4	0.0143	0.011233	1.273	0.215
x_1^2	0.020579	0.016815	1.224	0.233
x_3^2	0.046958	0.014501	3.238	0.004
x_1x_2	-0.007152	0.011768	-0.608	0.549
x_1x_3	-0.024825	0.013757	-1.805	0.084
x_1x_4	-0.024585	0.011768	-2.089	0.047
x_2x_3	0.035625	0.013757	2.59	0.016
x_3x_4	0.011025	0.013757	0.801	0.431

Table 5. Variance analysis of surface roughness in parallel direction

Source	DF	Seq. SS	Adj SS	Adj. MS	F	P
Regression	11	0.245556	0.245556	0.022323	13.27	0
Linear	4	0.206459	0.206459	0.051615	30.68	0
Square	2	0.018352	0.020161	0.01008	5.99	0.008
Interaction	5	0.020745	0.020745	0.004149	2.47	0.061
Residual error	24	0.040374	0.040374	0.001682		
Total	35	0.285931				

$$\begin{aligned} Ra = 0{:}544944 \quad & 0{:}063875x_1 + 0{:}199292x_3 \\ & + 0{:}072131x^2{}_1 \quad 0{:}049802x_1x_2 \quad 0{:}080341x_1x_4 \end{aligned} \tag{2}$$

$$\begin{aligned} Ra = 0{:}210528 & + 0{:}03695x_2 + 0{:}082792x_3 \\ & + 0{:}0{:}0469589x^2{}_3 \quad 0{:}024585x_1x_4 \quad 0{:}035625x_2x_3 \end{aligned} \tag{3}$$

Using MATLAB environment, the GA optimization is carried out. By varying GA parameters, the different optimized process parameters were obtained and the resulted parameters further used for optimization in the following manner:

Population type: Double vector; Population size: 100; Number of generations: 200; Number of stall generation: 50; Fitness function: Rank scaling; Selection function: Roulette wheel, Crossover function: Two point; Crossover fraction: 0.8; Mutation function: adaptive feasible; Migration: Forward, Migration fraction: 0.2.

The condition of optimal material removal, offering least surface roughness is shown in Table.4. An experiment was performed based upon the optimal parameter settings for surface roughness to achieve the targeted surface finish. Table.4 demonstrated the average surface roughness (Ra) by regression model and the experimental Ra with the setting of optimal parametric setting as acquired from the GA. The result revealed, a reliable prediction with reasonable accuracy as the percentage error of the predicted average surface roughness value is in close correlation with the experimentally observed value.

Table 6. The optimum value of the process parameters

Response	XY speed (mm/s)	Z speed (mm/s)	Spindle speed (rpm)	Depth of cut (mm)	Predicted Value	Exp. Value	% error
Ra in perpendicular direction	4	7	8000	0.1	0.2048	0.314	34.777
Ra in parallel direction	4	7	8000	0.1	0.1339	0.138	2.971

7. CONCLUSIONS

In the present study, the roughness (experimental results) obtained from the extensive machining conducted on the surface of green alumina compact with diverse input parameters of CNC machine using diamond embedded tool are compared and validated with the predictions. A hybrid GA based RSM approach was successfully projected to obtain optimum CNC machining parameters for the prediction of optimum surface roughness of CNC machined green alumina surface in both the directions of tool movements (along and perpendicular direction of tool path movement). Initially, the RSM model was employed to build up the mapping between input process parameters of CNC milling machine and surface roughness. Further, the developed regression models were combined with the projected GA to acquire optimum material removal parameters that leading to the minimum surface roughness value. It was observed that prediction accuracy of hybrid GA based RSM were fairly close correlation with the experimental values with maximum percentage absolute error of 34.777 and 2.971, in perpendicular and parallel direction of tool movement, respectively. Thus, the obtained hybrid model can be an ideal approach to level the input parameters of CNC Milling of ceramic compacts with reasonable accuracy for prediction of surface finish, and hence a significant saving of cost and time.

ACKNOWLEDGEMENTS

The author acknowledges BMTE Lab, School of Medical Science and Technology, IIT Kharagpur-721302 for assistance with the development, CNC machining and characterization of ceramic compacts.

IOP Conf. Series: Materials Science and Engineering **410** (2018) 012022 doi:10.1088/1757-899X/410/1/012022

REFERENCES

[1] L.M. Sheppard, 1991, Ceramic Bulletin, Vol. 70, pp. 692.

[2] R.W. Rice, 1990, Advances in Chemical Engineering Journal, Vol. 36, pp. 481.

[3] M. Savitz, 1999 , American Ceramic Society Bulletin, Vol. 78, pp. 53.

[4] J. Yang, J. Yu, Y. Cui, and Y. Huang, 2012, Ceramics International, Vol. 38, pp. 3643.

[5] P. Dadhish, P.K. Srivas, S. Mohanty, S. Dhara, 2015, Journal of European Ceramic Society, Vol.35, pp. 3909.

[6] S. Baskaran, G.D. Maupin, G.L. Gordon, 1998, American Ceramic Society Bulletin, Vol. 77 pp 53

[7] J.D. Cawley, 1999, Current Opinion in Solid State and Materials Science, Vol. 4, pp. 483.

[8] G.P Birkby, P Dranseld, J.H McColgan, J.R.G. Evans, 1994, British Ceramic Transection, Vol. 93, pp. 183.

[9] S. Dhara and B. Su, 2005, International Journal of Applied Ceramic Technology, Vol. 2, pp. 262.

[10] B. Su, S. Dhara, and L. Wang, 2008, Journal of the European Ceramic Society, Vol. 28, pp. 2109.

[11] R.K. Kamboj, S. Dhara, and P. Bhargava, 2003, Journal of the European Ceramic Society, Vol. 23, pp. 1005.

[12] S. Mohanty, A.P. Rameshbabu, and S. Dhara, 2013, Ceramic International, Vol. 39, pp. 8985.

[13] M.S, Lou. J.C Chen., C.M Li., 1998, Industrial Technology, Vol.15, pp. 1.

[14] P.G. Bernados and G.C. Vosniakos, 2003, International Journal of Machine Tools and Manufacture, Vol. 43. No.8, pp. 833.

[15] J.L, Yang. J.C, Chen, 2001, Industrial Technology, Vol. 17, pp. 1.

[16] Pradhan, Das R, Biswas, 2009 International Journal on Design and Manufacturing Technologies, Vol. 2, pp. 20.

[17] S. Zainal, K. Z. Nadzirah*, A. Noriham, I. Normah, 2013, Agricultural Sciences, Vol. 5B, pp. 65.

[18] P. V. S. Suresh, P. V. Rao, and S. G. Deshmukh, 2002, International Journal of Machine Tools and Manufacture, Vol. 6, pp. 675.

[19] N. V. Prajina, 2013, International Journal of Engineering Research and Technology, Vol. 6, pp. 739.

[20] B. C. Routara, A. Bandyopadhyay, and P. Sahoo, 2009, International Journal of Advanced Manufacturing Technology, Vol. 11, pp. 1166.

[21] H. Oktem, T. Erzurumlu F. Erzincanli, 2006, Materials Design, Vol. 9, pp. 735.

[22] S Mohanty, B Das, S Dhara, 2013, Journal of Asian Ceramic Societies, Vol. 2, pp. 184.

[23] S Mohanty, A Rameshbabu, S Mandal, B Su, S Dhara 2013, Journal of Asian Ceramic Societies, Vol. 1, pp. 274.

IOP Conf. Series: Materials Science and Engineering **410** (2018) 012023 doi:10.1088/1757-899X/410/1/012023

Prediction of flexural Strength of SFRC form Concrete Strength without fibres

A. M. Shende *[1], A.M. Pande[2] and Sakshi Manchalwar[1]

[1]Priyadarshini J.L. College of Engineering, Nagpur, Maharashtra, India.
[2]Yeshwantrao Chavan College of Engineering, Nagpur, Maharashtra, India.
Corroseponding Email address: shende_renu@yahoo.com

ABSTRACT

Results of an investigation conducted to study the Flexural strength of steel fiber reinforced concrete (SFRC) containing fibers of 1%, 2% and 3% volume fraction of Hook tain steel fibers of 50, 60 and 67 aspect ratio are presented. Beam specimens of dimension 100 x 100 x 500 mm were tested under compression testing machine as per I.S. 516-1959. A result data obtained is analyzed and compared with control concrete specimen without fibers. From the data from various references and own data a mathematical models is developed which predicts flexural strength with the help of control concrete strength without fibers, percentage of fibers (Vf) and aspect ratio (AR). The mathematical model is found to be very successful.

Key words: Concrete, Fibers, Aspect ratio, Flexural Strength

1. INTRODUCTION

Concrete is most widely used construction material in the world. Because of its ability to get cast in any form and shape, it has almost replaced old construction materials such as brick and stone masonry. The strength and durability of concrete can be changed by making appropriate changes in its ingredients like cementitious material, aggregate and water and by adding some special ingredients. Hence concrete is very well suited for a wide range of applications. However, concrete has some deficiencies as listed below

1) Low tensile strength
2) Low post cracking capacity
3) Brittleness and low ductility
4) Limited fatigue life
5) Not capable of accommodating large deformations
6) Low impact strength

The presence of micro cracks at the mortar-aggregate interface is responsible for the inherent weakness of plain concrete. The weakness can be removed by inclusion of fibres in the mix. Different types of fibers such as those used in traditional composite materials have been introduced into the concrete mixture to increase its toughness, or ability to resist crack growth. Thus the fibre-reinforced concrete is a composite material essentially consisting of conventional concrete or mortar reinforced by fine fibres.

The fibres can be imagined as an aggregate with an extreme deviation in shape from the rounded smooth aggregate. The fibres interlock and entangle around aggregate particles and considerably reduce the workability, while the mix becomes more cohesive and less prone to segregation. The fibres are dispersed and distributed randomly in the concrete during mixing, and thus improve concrete properties in all directions. Fibres help to improve the compressive strength, split tensile strength, flexural strength, post peak ductility performance, pre-crack tensile strength, fatigue strength, impact strength and eliminate temperature and shrinkage cracks. Essentially, fibres act as crack arrester restricting the development of cracks and thus transforming an inherently brittle matrix, i.e. cement concrete with its low tensile and impact resistances, into a strong composite with superior crack resistance, improved ductility and distinctive post-cracking behaviour prior to failure. Hence this study explores the feasibility of steel fiber reinforcement; aim is to do parametric study on compressive strength, flexural strength, tensile strength study etc. with variables of grade of concrete, aspect ratio and percentage of steel.

2. EXPERIMENTAL PROGRAMME

2.1.Material used

Cement: Ordinary Portland cement of 53 grade was used in this experimentation conforming to I.S. – 12269- 1987.

Sand: Locally available sand zone II with specific gravity 2.65, water absorption 2% and fineness modulus 2.92, conforming to I.S. – 383-1970.

Coarse aggregate: Crushed granite stones of 10 mm size having specific gravity of 2.70, fineness modulus of 2.73, conforming to IS 383-1970

Water: Potable water was used for the experimentation.

Superplasticizer: To impart additional workability a superplasticizer (Rheobuild 1100) 0.6 % to 0.8% by weight of cement was used. It is based on sulphonated naphthalene polymers with following properties as per I.S. – 9103-1999.

Fibers:

Steel Fibers: - In this experimentation Hook tain Steel fibres were used. The different aspect ratios adopted were 50, 60, and 67.

IOP Conf. Series: Materials Science and Engineering **410** (2018) 012023 doi:10.1088/1757-899X/410/1/012023

Table 1. Geometrical properties of steel fiber.

Aspect Ratio	Length (mm)	Diameter (mm)
50	35	0.70
60	30	0.50
67	30	0.40

3. EXPERIMENTAL METHODOLOGY

For flexural strength test beam specimens of dimension 100 x 100 x 500 mm were cast. The specimens were demoulded after 24 hours of casting and were transferred to curing tank where they allowed curing for 28 days as per IS 516-1970. These specimens were tested under two-point loading as per IS 516-1959, over an effective span of 400 mm on Flexural testing machine. Load and corresponding deflections were noted up to failure. In each category three beams were tested and their average value is reported. The flexural strength was calculated as follows.

Fl}}5}}d}}}}}}}g}h(MP}})=PL...........(1)

Where,

P = Failure load,

L = Center to center distance between the support = 400 mm,

b = width of specimen, 100 mm

d = depth of specimen, 100 mm.

Fig. 1. Testing of Flexural strength test specimen.

IOP Conf. Series: Materials Science and Engineering **410** (2018) 012023 doi:10.1088/1757-899X/410/1/012023

3.1. MATHEMATICAL MODEL FOR FLEXURAL STRENGTH

To arrive at mathematical model for prediction of various strength parameters, the process started with development of some elementary mathematical relations and then arriving at some generalized equations.

3.1.1Predicted Compressive Strength for M-20 Grade:-Preliminary mathematical equation for prediction of compressive strength the equation was developed with understanding of development of strength. The first mathematical equation is presented below, other equations are developed using similar approach.

Table 3.1 – Compressive Strength of SFRC with 0% fibres for M20 grade

Compressive strength (MPa)	Average Compre ssive strength (MPa)
32 32 31.56	31.85

Table 3.2 – Compressive Strength of SFRC with 1%, 2% and 3% fibres

Different aspect ratios of fibres	For SFRC with 1% fibres	For SFRC with 2% fibres	For SFRC with 3% fibres
	Average Comp. strength (MPa)	Average Comp. strength (MPa)	Average Comp. strength (MPa)
50	37.48	37.93	39.26
60	36.44	37.19	38.22
67	34.89	36.00	37.11

Mathematical equation for compressive strength prediction:

Compressive strength of SFRC (σ_{SC}), using multiplying factor (F) as follows

$\sigma_{SC} = F \text{ x } \sigma_{mc}$ ____________________ (3.1)

Equation 6.1 can be written for 1%, 2% and 3% fibres, using multiplying factors F_1, F_2 and F_3 as $\sigma_{SC} = 1F_1 \text{x } \sigma_{mc}$, and $\sigma_{SC} = 2F_2 \text{ x } \sigma_{mc}$ and $\sigma_{SC} = 3F_3 \text{ x } \sigma_{mc}$ from which it can be written

$F_1 = \sigma_{SC} /1 \text{ x } \sigma_{mc}$ ________________ (3.2)

$F_2 = \sigma_{SC} /2 \text{ x } \sigma_{mc}$ ________________ (3.3)

$F_3 = \sigma_{SC} /3 \text{ x } \sigma_{mc}$ ________________ (3.4)

Using experimentally observed compressive strength values, factors F_1, F_2 and F_3 are tabulated as follows.

Table 3.3 – Multiplying factor for compressive strength for M20 grade

Aspect Ratio	F1 (for 1% fibres)	F2 (for 2% fibres)	F3 (for 3% fibres)
0	1	0.5	0.33
50	1.17	0.41	0.27
60	1.14	0.58	0.4
67	1.09	0.56	0.38

Using statistical approach, the best fit exponential equations are obtained for multiplying factors F_1, F_2 and F_3, keeping them as dependent variables on aspect ratios (A)

$F_1 = 1.0145\ e^{0.001786A}$ ______ (3.5) ___ for 1% fibres

$F_2 = 0.4781\ e^{0.001368A}$ ______ (3.6) ___ for 2% fibres

$F_3 = 0.31441\ e^{0.001847A}$ ______ (3.7) ___ for 3% fibres

Factors F1, F2 and F3, can be used to predict compressive strength for SFRC with 1%, 2% and 3% fibres respectively. Now equation 3.1 can be written as

$\sigma_{SC} = \sigma_{mc} \times P\ e^{QA}$ ______(3.8)

Comparing equation 3.5, 3.6, 3.7 and 3.8 it is seen that the variables P and Q depend on fibre percentage. Using equation 3.5, 3.6, 3.7 and statistical approach, the best fit polynomial percentages of fibres (V_f) as follows.

Now rearranging equation 3.8 and putting V_f for percentage of fibres it can be written

$\sigma_{SC} = \sigma_{mc} \times V_f \times P\ e^{QA}$ ______(3.9)

Table 3.4 – Predicted and Observed Compressive Strength of SFRC for M-20

Different aspect ratios of fibers	For SFRC with 1% fibers			For SFRC with 2% fibers			For SFRC with 3% fibers		
	Predicted Comp. Strength (MPa)	Observed Comp. Strength (MPa)	Ratio of Predicted to observed comp. strength	Predicted Comp. Strength (MPa)	Observed Comp. Strength (MPa)	Ratio of Predicted to observed comp. strength	Predicted Comp. Strength (MPa)	Observed Comp. Strength (MPa)	Ratio of Predicted to observed comp. strength
0	32.31	31.85	1.014	30.45	31.85	0.956	30.04	31.85	0.943
50	35.33	37.48	0.943	32.61	37.92	0.860	32.95	39.25	0.839
60	35.97	36.44	0.987	33.06	37.185	0.889	33.56	38.22	0.878
67	36.42	34.89	1.044	33.38	36.00	0.927	34	37.11	0.916

IOP Conf. Series: Materials Science and Engineering **410** (2018) 012023 doi:10.1088/1757-899X/410/1/012023

The above model can be used to predict compressive strength of SFRC by putting appropriate values of percentage of fibers (V_f) and aspect ratio (A). The values of P and Q are given by equation 3.5 and 3.7

Following table gives predicted and observed compressive strengths of SFRC along with a ratio of predicted compressive strength to observed compressive strength, for different aspect ratios of fibers.

Where,

σ_{SC} = compressive strength of SFRC

σ_{mC} = compressive strength of plain concrete

V_f = percentage of fibers

A = aspect ratio of fiber

3.2 Similar Mathematical models are developed to predict the strength properties of Steel fibre reinforced concrete for M20, M30 and M40 grade of concrete, for aspect ratio 50, 60 and 67 and percentage of steel 1%, 2% and 3% the details are given below

Table 3.5 – Values of P and Q by using mathematical model

Model No.	Predicted Parameter	Formula	Grade of Concrete	% Fibers (V_f)	Value of P	Value of Q
A	28 days Comp. Strength	$\sigma_{SC} = \sigma_{mc}$ x V_f x P e^{QA}	M-20	1	1.0145	0.001786
B	28 days Comp. Strength	$\sigma_{SC} = \sigma_{mc}$ x V_f x P e^{QA}	M-20	2	0.4781	0.001368
C	28 days Comp. Strength	$\sigma_{SC} = \sigma_{mc}$ x V_f x P e^{QA}	M-20	3	0.31441	0.001847

Mathematical model (M1) is developed using mathematical equations developed in earlier sections and trend lines observed in the graphs.

Mathematical Model (M1) is developed to predict strength of SFRC by using strength of controlled concrete (concrete without fibers), hence after referred as Control Strength in the models, percentage of fibers and Aspect ratio (A.R.). Same mathematical model can predict compressive strength, flexural strength and split tensile strength

By using data from references and own data (total 128 observations) mathematical model is developed, it is presented as given below:

SFRC Strength = Control Strength * EXP ((0.893*LN (% Fibre) +1.954)/1000*AR)

Where SFRC Strength- flexural strength of SFRC, Control Strength-flexural strength of concrete without fibres, % Fibre- percentage of fibres in concrete mix and AR-aspect ratio of steel fibres.

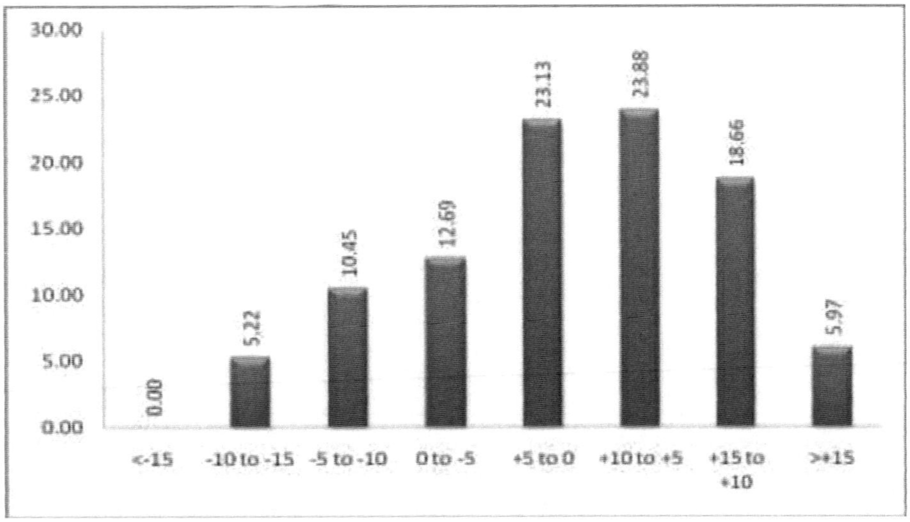

Fig. 2. Reliability of Mathematical for Flexural Strength

Flexural strength values for other research papers and own research flexural strength values along with predicted flexural strengths using above equation are presented in table 3. Reliability of the model is presented in graph given below:

Table 2. Flexural strength of SFRC

Different aspect ratios of fibres	For SFRC with 1% fibres		For SFRC with 2% fibres		For SFRC with 3% fibres	
	Flexural strength (MPa)	Average Flexural strength (MPa)	Flexural strength (MPa)	Average Flexural strength (MPa)	Flexural strength (MPa)	Average Flexural strength (MPa)
50	8.27 8.8 8.8	8.62	9.33 10 9.47	9.6	10.27 11.33 10.40	10.66
60	8.07 8.67 8.40	8.38	9.20 9.33 9.20	9.24	9.73 9.87 10	9.88
67	8.00 8.53 8.27	8.26	9.13 9.20 9.00	9.11	9.40 9.60 9.73	9.57

Table 3. Flexural strength values from own and other literature and their predicted Flexural strengths by using mathematical model

1	2	3	4	5	6	7	8
SN	Ref. No.	Control Strength	% of Fibre	Aspect Ratio	Flexural Strength	Predicted Flexural Strength	% Error
1	45	5.65	1	60	6.35	6.35	-0.04
2	45	5.65	1.5	60	6.87	6.49	5.50
3	45	5.65	2	60	7.19	6.59	8.30
4	45	5.65	0.5	60	5.43	6.12	-12.73
5	45	5.65	0.25	60	5.39	5.90	-9.43
6	45	5.65	1.5	60	6.02	6.49	-7.85
7	45	5.65	1	60	6.56	6.35	3.16
8	45	5.65	0.5	60	6.44	6.12	4.95
9	42	6.9	0.5	55	6.8	7.43	-9.20
10	42	6.9	0.75	55	7	7.58	-8.22
11	42	6.9	1	55	7.5	7.68	-2.44
12	42	6.9	0.5	55	7.3	7.43	-1.72
13	42	6.9	0.75	55	8.2	7.58	7.62
14	42	6.9	1	55	6.8	7.68	-12.98
15	53	6.13	0.25	50	6.13	6.35	-3.65
16	53	6.13	0.5	50	6.13	6.55	-6.90
17	53	6.13	0.75	50	6.6	6.67	-1.10
18	53	6.13	1	50	6.87	6.76	1.61
19	37	5.04	1	38	5.52	5.43	1.66
20	37	5.04	1	38	5	5.43	-8.57
21	37	5.04	1	38	6.3	5.43	13.83
22	37	5.04	1	38	5.64	5.43	3.75
23	37	5.04	1	38	5.04	5.43	-7.71
24	37	5.04	1	38	5.64	5.43	3.75
25	37	5.04	1	38	5.28	5.43	-2.81
26	37	5.04	1	38	5.7	5.43	4.76
27	37	5.04	1	38	5.64	5.43	3.75
28	99	7.24	0.5	50	7.87	7.74	1.65
29	99	7.24	1	50	8.17	7.98	2.29
30	99	7.24	1.5	50	9.13	8.13	10.96
31	99	7.24	2	50	8.91	8.23	7.59
32	99	7.24	2.5	50	8.67	8.32	4.08
33	99	7.24	3	50	8.44	8.38	0.66

34	99	7.24	3.5	50	8.11	8.44	-4.10
35	56	3.73	0.5	85	4.68	4.18	10.72
36	56	3.73	1	85	4.74	4.40	7.09
37	56	3.73	1.5	85	4.86	4.54	6.55
38	56	3.73	2	85	5.27	4.64	11.92
39	56	3.73	2.5	85	5.39	4.72	12.41
40	56	3.73	3	85	5.51	4.79	13.12
41	56	3.73	3.5	85	5.63	4.84	13.97
42	50	6.52	0.5	80	7.04	7.25	-3.05
43	50	6.52	1	80	7.64	7.62	0.22
44	50	6.52	1.5	80	7.98	7.85	1.66
45	50	6.52	2	80	8.51	8.01	5.87
46	50	6.52	2.5	80	8.3	8.14	1.94
47	50	6.52	3	80	8.02	8.25	-2.81
48	50	6.52	3.5	80	7.75	8.34	-7.57
49	85	6.21	0.5	80	7.15	6.91	3.36
50	85	6.21	1	80	7.73	7.26	6.07
51	85	6.21	1.5	80	8.13	7.47	8.07
52	85	6.75	0.5	80	8.06	7.51	6.81
53	85	6.75	1	80	8.54	7.89	7.59
54	85	6.75	1.5	80	9.15	8.12	11.21
55	85	7.4	0.5	80	8.76	8.23	6.00
56	85	7.4	1	80	9.32	8.65	7.17
57	85	7.4	1.5	80	10.13	8.91	12.08
58	85	8.02	0.5	80	9.58	8.92	6.85
59	85	8.02	1	80	10.36	9.38	9.49
60	85	8.02	1.5	80	11.01	9.65	12.33
61	92	4.91	0.8	50	4.68	5.36	-14.54
62	92	4.91	0.8	50	5.07	5.36	-5.72
63	92	4.91	0.8	50	4.69	5.36	-14.29
64	26	7.06	0.5	60	7.16	7.65	-6.83
65	26	6.04	1	55	6	6.73	-12.09
66	Own	7.2	1	50	8.4	7.94	5.49
67	data	7.2	1	50	8.8	7.94	9.78
68	0	6.8	1	50	7.6	7.50	1.34
69	0	7.2	1	60	8.2	8.10	1.27
70	0	7.2	1	60	7.6	8.10	-6.52
71	0	6.8	1	60	8.4	7.65	8.98
72	0	7.2	1	67	8	8.21	-2.59
73	0	7.2	1	67	8.4	8.21	2.30
74	0	6.8	1	67	7.6	7.75	-1.99
75	0	7.2	1	50	8	7.94	0.76

IOP Conf. Series: Materials Science and Engineering **410** (2018) 012023 doi:10.1088/1757-899X/410/1/012023

76	0	7.6	1	50	8.8	8.38	4.77
77	0	6.8	1	50	9.6	7.50	21.90
78	0	7.2	1	60	9.2	8.10	12.00
79	0	7.6	1	60	7.6	8.55	-12.44
80	0	6.8	1	60	9.2	7.65	16.89
81	0	7.2	1	67	8.4	8.21	2.30
82	0	7.6	1	67	8.8	8.66	1.56
83	0	6.8	1	67	8.4	7.75	7.72
85	0	7.6	1	50	8.8	8.38	4.77
86	0	7.2	1	50	9.2	7.94	13.71
87	0	7.6	1	50	8.4	8.38	0.24
88	0	7.6	1	60	8.4	8.55	-1.73
89	0	7.2	1	60	8.8	8.10	8.00
90	0	7.6	1	60	8	8.55	-6.82
91	0	7.6	1	67	8	8.66	-8.29
92	0	7.2	1	67	8	8.21	-2.59
93	0	7.6	1	67	8.8	8.66	1.56
94	0	7.2	2	50	8.8	8.19	6.95
95	0	7.2	2	50	9.6	8.19	14.70
96	0	6.8	2	50	9.6	7.73	19.44
97	0	7.2	2	60	8.4	8.40	-0.02
98	0	7.2	2	60	9.2	8.40	8.67
99	0	6.8	2	60	10	7.94	20.65
100	0	7.2	2	67	8.4	8.55	-1.84
101	0	7.2	2	67	9.6	8.55	10.89
102	0	6.8	2	67	9.4	8.08	14.05
103	0	6.8	2	50	8.8	7.73	12.12
104	0	7.2	2	60	8.8	8.40	4.52
105	0	7.6	2	60	10	8.87	11.31
106	0	6.8	2	60	9.2	7.94	13.75
107	0	7.2	2	67	8.4	8.55	-1.84
108	0	7.6	2	67	9.6	9.03	5.94
109	0	6.8	2	67	9.6	8.08	15.84
110	0	7.6	2	50	8.8	8.64	1.78
111	0	7.2	2	50	9.6	8.19	14.70
112	0	7.6	2	50	10	8.64	13.57
113	0	7.6	2	60	8.8	8.87	-0.78

IOP Conf. Series: Materials Science and Engineering **410** (2018) 012023 doi:10.1088/1757-899X/410/1/012023

114	0	7.2	2	60	9.2	8.40	8.67
115	0	7.6	2	60	9.6	8.87	7.62
116	0	7.6	2	67	8	9.03	-12.87
117	0	7.2	2	67	9	8.55	4.95
118	0	7.6	2	67	10	9.03	9.70
119	0	7.2	3	60	9.2	8.59	6.67
120	0	7.2	3	60	9.6	8.59	10.56
121	0	6.8	3	60	10.4	8.11	22.02
122	0	7.2	3	67	8.8	8.76	0.40
123	0	7.2	3	67	9.4	8.76	6.76
124	0	7.2	3	60	9.2	8.59	6.67
125	0	7.6	3	60	9.6	9.06	5.59
126	0	6.8	3	60	10.8	8.11	24.91
127	0	7.2	3	67	10	8.76	12.35
128	0	7.6	3	67	9.6	9.25	3.63
129	0	6.8	3	67	9.2	8.28	10.02
130	0	7.6	3	60	9.6	9.06	5.59
131	0	7.2	3	60	10	8.59	14.14
132	0	7.6	3	60	10.4	9.06	12.85
133	0	7.6	3	67	8.8	9.25	-5.13
134	0	7.2	3	67	10.4	8.76	15.72
135	0	7.6	3	67	10	9.25	7.48

4. CONCLUSIONS

The following conclusions could be drawn from the present investigation.

1. It is observed that flexural strength is on higher side for 3% fibres as compared to that produced from 0%, 1% and 2% fibres.
2. The flexural strength properties are observed to be on higher side for aspect ratio of 50 as compared to those for aspect ratio 60 and 67.
3. The failure pattern of specimen reveals improved ductility of SFRC due to the addition of steel fibers over concrete without steel fibres.
4. Mathematical model developed from control strength, aspect ratio and percentage of fibre is found to be valid to over 85% confidence limit.

5. REFERENCES

[1] N. Ganeshan, 2007, Civil Engineering and construction Review, Vol. xxx, pp.54.
[2] V. Bhikshma, K. Ravande, and K. Nitturkar, 2005, Recent advances in concrete and construction, Chennai, Vol. xxx, pp 23.
[3] N.R. Murthy Dakshina, International conference recent advance on concrete and construction technology, 2005.

[4] P. Balaguru and H. Najm, 2004, Material Journal, Vol. 101, pp. 281.
[5] D. Tensing, L. Jeminah, and S. Jaygopal, National seminar on advance in construction materials, 2003.
[6] R. M. Damgir and M.I.M. Ishaque, Proceeding of the INCONTEST 2003, Coimbatore, 2003, pp 281.
[7] P.S. Raghuprasad and Ravindranatha, Proceeding of the INCONTEST 2003, Coimbatore, 2003, pp 403.
[8] J. Permalatha and V. Govindraj, Proceeding of the INCONTEST 2003, Coimbatore, 2003, pp. 462.
[9] Z. Bayasi and J. Zeng, ACI Materials Journal, 90 (1993), pp 605.
[10] P. Soroushian and Bayasi, 1991, ACI Materials Journal, Vol. 88, pp 129.
[11] A.S. Ezeldin and S.R. Lowe, 1991, ACI Materials Journal, Vol. 88, pp 384.
[12] V. Ramakrishnan, G.Y. Wu, and G. Hosalli, 1989, Transportation Research Record, Vol. 1226, pp 69.
[13] BIS 383:1970 Specification for coarse and fine aggregates from natural sources for concrete (second revision).
[14] BIS 10262:2007 Recommended guidelines for concrete mix design.
[15] BIS 1727:1967 (Reaffirmed 2004) Methods of test for pozzolanic materials (first revision)
[16] BIS 4031:1988 (Reaffirmed 2000) Methods of physical tests for hydraulic cement
[17] Jagannathan, Proceeding of International conference on Innovation, 2010, pp 347.
[18] A. H. Saif, Y. Nur, and T. Kamal, Proceedings of ICFRC, 2004, pp 135.
[19] D. A. Hari and M. Srikanth, Proceedings of ICFRC, 2004, pp 167.
[20] V. Kumar, A.K. Sinha, and M.M. Prasad, Proceedings of ICFRC, 2004, pp 177.
[21] C. Sashidhar, H. Sudharsan Rao and N.V. Ramana, Proceedings of ICFRC, 2004, pp 247.
[22] Y.M. Ghugal, C.V. Nandanwar, and S.S. Bansode, International conference on Innovation, 2010, pp 367.
[23] M. Sivaraja and S. Deepak, International conference on Innovation, 2010, pp 417.
[24] R.M. Sawant and Y.M. Ghugal, International conference on Innovation, 2010, pp 438.
[25] V.M. Lad, Y.M. Ghugal, and U.P. Waghe, International conference on Innovation, 2010, pp. 446.
[26] S. Eswari, P.N. Raghunath, and S. Kothandaraman, 2011, ARPN Journal of Engineering and Applied Sciences, Vol. 6, pp 1.
[27] A.S. Ezeldin and S.R. Lowe, 1991, ACI Materials Journal, Vol. 88, pp 384.
[28] G. Marciukaitis, R. Salna, and B. Jonaitis, 2011, Journal of Civil Engineering and Management, Vol. 17, pp 137.

Institute of Physics
Dirac House, Temple Back
Bristol BS1 6BE UK

ISSN: 1757-8981
ISBN 978-1-5108-7195-3

9 781510 871953

CURRAN ASSOCIATES INC.
proceedings.com